W9-CAO-004

INSPIRE

ALSO BY BEN FOGLE

Up
English
Land Rover
Labrador
The Accidental Naturalist
The Accidental Adventurer

BEN FOGLE

INSPIRE

LIFE LESSONS FROM THE WILDERNESS

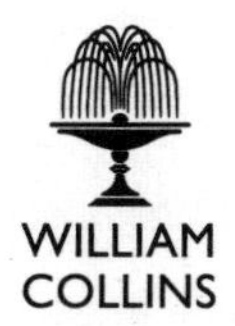

WILLIAM
COLLINS

William Collins
An imprint of HarperCollins*Publishers*
1 London Bridge Street
London SE1 9GF

WilliamCollinsBooks.com

First published in Great Britain by William Collins in 2020

2021 2023 2022 2020
2 4 6 8 10 9 7 5 3 1

A catalogue record for this book is
available from the British Library

ISBN 978-0-00-837403-7 (hardback)
ISBN 978-0-00-837404-4 (trade paperback)

Typeset in Minion Pro
Printed and bound in Great Britain by
CPI Group (UK) Ltd, Croydon

To *you*

I hope that this book will help you find confidence,
hope and happiness

CONTENTS

PRELUDE

Spring 2020

A treehouse in Oxfordshire

Dear Reader,
I always had a plan.

In the event of a global crisis, I would retreat to the wilderness, an island in Scotland perhaps, where I would cultivate my own produce and home-school my children from our idyllic little croft far away from the crisis.

And now, the unimaginable has happened and I find myself in lockdown, isolated from the isolation of the Western Isles. Instead I am sitting in my new office, the children's treehouse at home, which has been requisitioned for the duration.

Like so many people, I have fought my own mental and physical battles in isolation. Keeping the dark clouds away before they have a chance to form into a storm. And also taking this time to become more introspective. Away from the bombardment of ever-changing information, lockdown has given me the opportunity to reflect and scrutinise what I have done, what I am doing and where I'm going.

All my life I have been running from an inner voice of doubt, the loud voice of failure and disaster that goads me and taunts me with its negativity and hopelessness. I have spent a lifetime trying to silence the pessimism by taking myself out of my comfort zone and out into nature, where I get perspective and direction.

Sometimes it takes a jolt like a world crisis to shock us from the slumber of complacency and remind us what is really important in life.

Many of us spend our lives searching for something that is often right under our noses all along … a sense of belonging. This sense of belonging brings contentment, inclusion and happiness that eludes many for their entire lives. They say home is where the heart is. The wilderness and nature is all our 'home' – a place where we can all belong.

For me, one of the most powerful forms of escape has been to think about the people, the places and landscapes of my life. I find myself disappearing from the chaos of the kitchen table, back to the mountains and forests, the jungles and islands, the open seas and polar ice caps, to rediscover the wilderness within.

There is so much we can learn about ourselves in the wilderness.

The wilderness forces us to rise to the challenge. To take responsibility for ourselves and adapt. We become more creative, less wasteful and more empathetic. In short, it makes us less selfish and more selfless.

Lockdown has given us all a period of retrospective reflection. What have we been doing? Where have we come from and where are we going? Without the background noise it has given us all an opportunity for contemplation. And I have been contemplating all the lessons I have learned from the wilderness, and the way my experiences in nature have shaped me.

This book was born from of the experiences of re-wilding myself. Written on my lap in the children's treehouse during the extraordinary period of lockdown isolation, here are lessons to inspire your own journey … and a sense of destination.

Ben

INTRODUCTION

'The wilderness holds answers to questions
we have not yet learned to ask.'
Nancy Newhall

This much I know.

It's easier to break something than it is to repair it. A falling china vase will smash into a thousand pieces in milliseconds. Of course it can be repaired, but those repairs require patience, skill and perseverance. It might take months or even years to carefully piece it back together, and even then you are left with a damaged vase, forever marked by the scars of repair.

Beauty of course is in the eye of the beholder. While some see only the scarring, others see beauty. It was the Japanese who invented the art of kintsugi, whereby porcelain repaired with gold becomes an art form and the finished product becomes even more valuable and covetable, each vein, line and repair representing hours of dedication and artistic craft. By embracing its flaws and imperfections, the item, a vase, for example, has appreciated in worth because of the trials and tribulations in repairing it.

Life is full of ups and downs, light and shade, peaks and troughs. In essence, we are all damaged vases, and the repairs build character and strength and spirit. We all seek healing in different places. Nothing is truly ever broken.

Throughout my life, the wilderness has been my healer. She is my friend and my mentor. I am not a religious person but, if I was, nature would be my church. My place of worship.

Nature has taught me respect and compassion. The wild teaches you the importance of valuing and rationing your resources, of caring for your environment, and of respecting nature. She has reminded me to be thankful for what I have rather than sorrowful for what I don't.

Throughout my life, through a series of adventures and misadventures, I have learned a great deal about myself through the theatre of nature. The wilderness has taught me how to deal with love and loss, with criticism and failure, with happiness and sorrow, with resilience and risk, aspiration and hope.

What goes up must come down. It's the reality of life. Summiting Everest marked a high point in my life on many different levels, but the reality was that I eventually had to come back down. I couldn't stay up there forever. Like a drug-induced high, there has to be a comedown.

In reality, my whole life has been a series of highs and lows. As a positive person I often skirt over the low points, but the year since climbing the highest mountain in the world has taught me a lesson about myself, about life and the importance of learning from our own failings and our weaknesses.

My childhood was dominated by failure: not always in obvious ways, but failings of one kind or another seemed to follow me like a bad smell. I failed my exams. I failed at friendships. I failed at sport. Deep down, this has scared me and it still scares me. I became failure and failure became me. It defined me as a person. I stopped trying because I knew I'd fail. There was no point. The last 45 years have been quite a rollercoaster. I don't want to sound too clichéd, but if I'm honest, my life has been defined by the pain of failure. The inevitability of this failure stripped me of confidence and self-esteem and crushed my spirit. And so, as a child, I found myself retreating to the trees and to the river bank and seashore as a form of therapy to try to rebuild that shattered confidence. As I grew older, the scale and scope of my relationship with nature and the wilderness would continue to grow and develop.

Growing up in the 1970s, my childhood failings were largely restricted to my own head. But 'failure' today is very different. In this social media-obsessed world of sharing, gloating, commenting, swiping and liking, failings are highlighted on a daily basis. Never have our lives been under such scrutiny while on display, held on a pedestal for all to see.

We are living in polarised times in which it is easier to find fault than to celebrate success. We are living the age of 'more', not just in terms of material consumption but in achievement, and even then we will pick holes and highlight the shortcomings. I don't know why we do it. Perhaps it is our unshakeable instinct towards jealousy? Maybe it's tall poppy syndrome. As the saying goes, 'Every time a friend succeeds, I die a little.'

So many people spend their lives worrying about what others are up to rather than enjoying their own lives; worrying more about comparing, contrasting and criticising than enjoying, fulfilling and respecting. But comparison kills happiness. It has always been the case, but rarely has it been so easy as it is in the age of the internet. Compare and contrast. Often we don't even realise that we are showing off. The human instinct seems to be to criticise and find failure.

And it is very easy to find fault. Throughout my life, however hard I've tried, people have always picked holes. When I was first chosen as a volunteer for the BBC social experiment *Castaway 2000*, in which I was marooned for a year on a remote island in the Outer Hebrides, 'it was only because I had a famous mother'; when I published my first book, 'it was only because I was a celebrity'; when I started presenting television, 'it was only because I was public school and posh'; when I presented the rural affairs programme *Countryfile*, I was dismissed as a townie who knew nothing of the countryside; when I presented the dog show Crufts, I was dismissed as knowing nothing about dogs; when I successfully rowed the Atlantic, 'it was only because we had the ocean currents to help us'; when I walked to the South Pole, 'it was only because we were supported'; when I climbed Everest it was dismissed by some because 'it was no longer a challenge'; and when I raise concerns about the environment, I am accused of gross hypocrisy because I fly.

I could go on and on … It seems black and white, and yet it isn't because you can make anything fit your own narrative with clever editing of the truth. Sometimes called 'fake news', it is more about using creative licence with the facts, or, as we used to call it, 'poetic licence'.

Never have our lives been more on show for all the world to see and criticise, presented not just to our family and friends but, depending on your use of social media, to millions of people, all with a voice and opinion of their own. The impact can be overwhelming and I'd be lying if I said it didn't affect me. Thin-skinned, I have always been stung by criticism. As much as I try to toughen my resolve to take it on the chin, I find myself recoiling from critique.

Throughout my life, through the criticism and the failure, I have had one consistent companion: the wilderness. Of course my family, my parents, my sisters, my wife, my children and my friends have always been there supporting me too, but the wilderness has always been my safe place.

It may seem a strange term to use, 'safe place', when the wilderness can be unpredictable and often hostile, but for me it has, in all of its many forms, from ocean to desert, jungle to forest, savannah to mountain, been a place of great comfort. It still surprises me how it is so often seen as a battle between man and wild, but for me, there has never really been a battle; it has always been a symbiotic relationship.

I love how the Swedes and the Norwegians refer to it as 'the nature', not just 'nature'. It's as if they are adding a reverence to the word to indicate its importance. For me, the wild, the nature, the wilderness, call it what you will, has been my medicine, a place to retreat from the pressures of daily life and a place to rebuild my confidence.

For the last eight years I have travelled the world meeting people who have dropped off the grid for my TV series *New Lives in the Wild*. Individuals who have suffered their own existential crisis and have decided to begin again with a simpler life in the wild. Each time I return from my ten-day immersion, I suffer my own mini crisis and ask questions about my own life. It's the old clichés, 'What's it all about?' and 'Why are we here?' They really are big questions with an infinite number of answers that wax and wane throughout life.

For many years I tried to confront my fear of failure. Rowing the Atlantic and climbing Mount Everest were just two examples in which I risked failure to overcome my own fear of it. If someone asks you to describe your life, chances are you will list your achievements and accomplishments and edit out the failings and shortcomings, but aren't they part of who we are? After all, failure is a part of everyday life. Without failure you can't really have success. It's all relative. In the same way that the millionaire wants to become a billionaire or the mayor wants to become the president, we will always aspire to more, and in doing so, we will have to face the inevitability of failures along the way. The problem is that most of us try to forget the failings. It's like we have amnesia.

Society treats failure like a taboo. But surely we are all a sum of our parts. Those experiences mould and shape us. If we learn from our mistakes then we become better people, don't we? If we only ever succeed without a struggle, then is it really success?

When you combine the self-doubt with the criticism levelled at any achievement by the pessimists and the naysayers, you get the perfect storm of doubt. I call them my 'downs'. I have always had them but they seemed to gather more often after my Everest trip. The small dark cloud that follows you even in sunny weather. I would be overcome with an often baseless anxiety, I began to worry more and overanalysed things.

In this picture-perfect Instagram world of Hollywood perfection, it is easy to edit our frailties and our fallibilities, but the reality is that we all have them, it's just that we don't often choose to share them. We all want to project the image of perfection, but the mask will eventually slip. I have always struggled with self-doubt. I think it might even be clinically described as having imposter syndrome. I lack self-worth. I feel like I have somehow cheated the system – after all, as a failure, I should have been destined to a life of failure. Success is not a part of who I am.

I wonder whether it is a symptom of the highly interconnected world in which we now live, where some people will be contrary for the sake of being contrary. Whatever the root cause, the result is that we have rarely lived under such intense scrutiny.

It is perhaps the reason that the wilderness has become increasingly precious as an antidote to the hate, pessimism and negativity that pervades modern life. Because in contrast the wilderness and nature offer a simple and honest world. The wilderness is free of complications. The term 'wild' may imply that it is dangerous and hostile, but it is arguably far safer than the urban world in which most of us choose to live.

Where the civilisation of humanity is often full of complication, nuance and confusion, the wilderness is an organic, simple place where black is black and green is green. There is no hate, or jealousy. There is no anger. Of course it can be dangerous and sometimes brutal. It can also be unpredictable, but we can learn a great deal from nature. I was certainly at my happiest in that environment.

Consider the happiness index. Is it really possible to be happy 100 per cent of the time? Think about it. We all aspire to it, but is it really possible? Surely happiness is created by fluctuating emotions? You need a small dip to recreate the emotion. I compare it to the weather. I don't know about you, but I have a smile on my face and experience an elevated mood whenever I wake to the morning sun streaming through the window. It feels like a treat, a rarefied moment on our cloudy isle.

I call it the 'California' syndrome. If you have ever been to California then you might recognise the reference. Each time I visit the west coast of America, I am seduced by its perfect weather. Clear blue skies and a perfect temperature seemingly *every* day. Every day the weather is guaranteed to be sunny. Can you imagine? It's the dream. The thing is, eventually it becomes boring. The sameness is repetitive and it loses its magic. We need the rainy days to really appreciate the sunny days.

A little like the weather, we need texture within our lives. We need to experience lows to really appreciate the highs. In short, without our failings we can never really taste the sweet smell of success. Our senses would become dulled and immune to it.

That is not to say that we need misery in order to attain happiness, but we need mixed emotions to feel real elation. Happiness must be the ultimate goal in life; no amount of money or success can ever really compete with happiness. So what is it that makes us *un*happy? Most young children are happy. As I write this book, my children are laughing and squealing and giggling and smiling. It's infectious, it makes me so happy to hear their happiness, I absorb it by osmosis.

If most young children are happy, then what is it that happens to us in life that makes us unhappy? If happiness is the default, then what happens to us to make us sad? There are plenty of people who have everything but feel like they have nothing. I am what I'd call an empath. I absorb the emotions around me. I feel other people's pain and happiness. It can be exhausting.

I would like to think of myself as a happy person. That's not to say I am always happy but I'd call myself a glass-half-full kind of guy. Physical things like sunshine, trees, water, mountains, grass, moss, flowers, rivers, lakes, ocean and boats all make me happy. And social things like family, walking, hiking, running, laughing, eating and chatting make me happy too. Combining the physical and the social would be my idyll.

When I was 13, my father took me and my best friend Toby on a summer camping trip to Algonquin Park in southeastern Ontario, Canada. We strapped the old cedar canoe that had been renovated by my grandfather to the roof of the Buick, packed our rucksacks with food and set off on our adventure.

The three of us slipped into the most remote wilderness I had ever encountered. During the daytime we paddled past elegantly grazing moose and wood-chipping beavers. At night, on pitching our tent we would hoist our bag of food high into the trees to keep it out of the reach of wild bears. I was terrified of bears.

(Before we left the UK, Dad had taken Toby and me to Hyde Park in central London for a bear-attack class. This consisted of Dad pretending to be a bear, battering us with his imaginary paws, while we rolled up in the foetal position to protect ourselves!) We used our fishing rods to catch freshwater fish for our dinner which we would supplement with freshwater clams that we dived for and wild blueberries collected from the forest. I can still remember our first campsite; the tent was only large enough for Toby and me, so Dad had to sleep outside.

I loved home but London never made me happy like Canada. In fact, Canada has always been my 'happy' place. The son of a Canadian, I spent my long summer holidays in Canada on the shores of Lake Chemong, a freshwater lake, in the wooden cabin hand-built by my grandfather. Those summers were a highlight of my childhood but also a catalyst for my love and appreciation of nature. For eight long weeks every year, my sisters, cousins and I swam, fished, paddled and camped around our little lake. I never remember feeling bored, and despite the frequent rain and plentiful mosquitoes, I never remember feeling uncomfortable. I loved the simplicity of lake life. We never wore shoes and rarely wore a shirt. My memories are of perpetual movement. We were always busy.

My late grandfather, Morris, was a bear of a man. He was a doer. A grafter. One of those people who can do, fix, mend or make anything. He had a number of different workshops around the cabin filled with tools and equipment.

As a child I never really appreciated the fact that he had built that cabin himself. Now I marvel at the fact that he did it. He hauled the timber and made the doors and he fitted the piping and wired the electrics. It was a bit wonky but everything worked and we loved it.

Have you ever been hit by a smell that takes you back in time? Well, for me, it is a slightly musky, damp smell that sends me reeling back to those heady summers on Lake Chemong. Uninsulated and without heating, the cabin would be boarded up by my grandfather each winter when the lake froze solid, before opening it once again in the spring, lifting its sagging corners with a car jack. The winter ice would have moved the docks and the boathouse, which he would single-handedly haul back into place.

While the lake wasn't exactly wilderness, it was our wild place where I could get lost in the flora and fauna. It was the antithesis of my London life. I spent hours in the canoe, paddling around the small islands that dotted the lake, watching the beavers. I wonder whether it is the reason I still feel such happiness and contentment when I am near water or in a forest; perhaps it is the subliminal memory of my childhood.

While I appreciate not everyone has a 'happy' childhood, over the years some of the happiest people I have met are those who have been able to recreate their childhood in later life; those carefree days without responsibility or the burden that comes with 'growing up'.

What is that burden? Is it responsibility? I have often wondered why, on a day-to-day basis, most childhoods have a higher return of happiness. Does age and responsibility hinder happiness? We all end up taking responsibility of one kind or another in later life, whether it is responsibility through work, family responsibilities or financial responsibilities. All of them come with a commitment and a pressure to provide. For many that 'pressure' can be overwhelming.

For me, the wild has always been an escape from the pressures of life. That is not to say there are no pressures in the wilderness, it's just that once they become familiar, they are less intimidating than the artificial pressures of urban life. When we talk about 'escapism' in the Western sense of the world, most people think of 'escaping' to the wilderness from the city, although there are of course still plenty of places in the world where people strive and dream of escaping rural life for the opportunity of the city.

It has always seemed strange to me that so many people dream of that escapism and yet so few actually do it. Escape? To escape implies that you are stuck somewhere that doesn't make you happy. Why then do so many of us choose to live a life without contentment, in the hope that we will somehow, somewhere along the line, find the elusive happiness? The interpretation is that we can only ever be happy at the beginning and end of our lives, the two periods when we are free of financial burden. Is it money then that makes us happy or unhappy? The fact that so many aspire to a simpler life in the wild suggests that most of us long for a life free from those money pressures.

When you live in a city, among people, materialism and money are king. Caught in the system, we need to accumulate money to keep up with the masses. Consumerism ties us into the financial grid and we become inextricably stuck.

The wilderness requires a very different kind of finance, that of respect, knowledge and patience. Life in the wild comes with great responsibility but it is a very different kind to that of consumerism.

The wilderness in all her forms, from the jungle to the ice caps and from the mighty rivers to the vast oceans, still makes up 23 per cent of the planet. But that means 77 per cent has been tamed and touched by man. There isn't much true wilderness left and our relationship with that sliver of real untouched nature remains complicated.

We like to take control of the flora and fauna all around us. We like to believe that we are the primary apex predator and that we have the power to tame, control and harness the landscape around us. Yet we also see the value of nature, and being out in the wild makes us happy.

I am happier outdoors than indoors. I am never more content than when I am walking or climbing, or hiking or running, or swimming. The wilderness lifts my spirits and boosts my confidence.

Over the years, through a series of expeditions, journeys, adventures and encounters, I have experienced a range of challenges that have tested my fortitude and resilience. They have inspired hope and happiness while other times eliciting despair and outrage. I have experienced courage and fear, blissful solitude and crippling loneliness. Success and failure in equal measure.

This is what I have learned along the way …

CHAPTER ONE

THE SEA

'The sea heals all ailments of man.'
Plato

Rowing across the Atlantic Ocean was the hardest thing I have ever done. The end.

This could have been a very short chapter, but I'll elaborate.

Rowing across the Atlantic was the hardest thing I have ever done. Really. I mean it. It tested me more than anything I have ever done before or will ever do again. Nothing will ever be as daunting, or boring, or as overwhelming, or mind-bendingly tedious.

It wasn't the scariest thing I have ever done, it wasn't the most dangerous thing I have ever done and it wasn't the toughest thing I have ever done, but it broke me. It also taught me about boredom, fear, pain, frustration, perseverance, planning and preparation. In fact, it would be one of my early lessons in the six 'P's:

Planning and Preparation Prevent Piss-Poor Performance. The ocean is almost unimaginably vast, a huge expanse of wilderness where we humans are particularly out of place. But in measuring myself against the Atlantic, I learned the value of endurance.

It had all started when an email popped into my Inbox inviting me to take part in the BBC's Sport Relief 2004 charity boxing match. I would be paired with Aled Jones, the young Welsh chorister who found fame as a teenager with a chart-topping cover of 'Walking in the Air' from *The Snowman*.

In an unusual and out-of-character move, I agreed. I should probably explain here that I am not a pugilist. I have always hated contact sport, and boxing had always seemed particularly unappealing. But sometimes you just go with your instincts. That split-second decision was like a tipping point for me, and the beginning of a great personal journey.

As it transpired, Aled Jones dropped out, and instead I was paired with the actor Sid Owen. For the first time in my life, I found myself well and truly out of my comfort zone, boxing on live national television in front of an audience of millions against a hugely popular *EastEnders* actor, who also happened to be a former amateur boxer.

The odds were stacked against me. At the weigh-in, Sid had chest-butted me, sending me flying backwards. 'IS THIS THE GREATEST SPORTING MISMATCH EVER?!' screamed the newspaper headlines the next day. I was the underdog. Not prepared to be humiliated, I threw myself into two months of hard training. I rose to the challenge and won the fight. In the end my greatest asset was my endurance. Committing 100 per cent to my training earned me the winner's belt.

After that, I was energised by survivor syndrome. I felt invincible. I felt that I could conquer the world. From there it was a sort of domino effect – the boxing bout was a catalyst for my increasing confidence in my own ability, and I soon found myself looking for another challenge that would really test my resilience and take me out of my comfort zone.

It was in that boxing ring, as my arm was lifted aloft and I was declared the winner, that the confidence to take on the Atlantic was kindled. Up to that point I was still a slave to the stereotyping of reality television, where my 'fame' had been created on BBC One's *Castaway 2000*. The thing about reality-show fame is that it is very shallow – it has neither substance nor roots. You are famous because you were on a television show, not because you have any depth, ability or knowledge. This was 2004 and I was still stuck in the *Heat* magazine celebrity circus and I wanted to escape from that. I craved for a challenge that would help me alter the course of my life and reclaim my own narrative.

I was still seen as just a posh and pretty public school boy. But I didn't forever want to be known as 'Ben, that guy from the reality TV show'. I wanted people to respect me for more than that. I wanted to surprise the critics who wrote me off and prove to them that beneath my puppy-like exterior was more grit and determination than they gave me credit for. If I could prove myself by suffering and enduring, then maybe I could take control of my public persona as well as build my own self-esteem and confidence.

Deep down I wanted success, not fame. To achieve that, I needed something big, while still within the realm of achievability. Something that would test me to my absolute limits. Rowing the Atlantic looked like the answer.

The first time I heard of anyone attempting to row across the Atlantic was when the husband-and-wife team of Andrew and Debra Veal announced their intention in 2001. There was a huge amount of media interest – a married couple, with the husband an amateur rower and the wife a complete novice, was bound to cause a stir. This only intensified when Andrew was forced to retire after suffering uncontrollable anxiety just eight days into the journey.

Debra took the decision to carry on alone. Like the rest of the nation, I was captivated by the unfolding drama. Debra had never rowed before she signed up for the race. I admired her resolve to carry on when her husband was rescued, and yet despite the temptation to abandon the row, she persevered and 108 days later reached land. I was enthralled by her story, and when it came to finding a challenge of my own, rowing the Atlantic seemed like a reasonable choice.

'Reasonable', of course, is different for each of us. As a child, I always loved boats. I spent a great deal of time in canoes and rowing boats at my grandparents' lakeside cabin, and while the ocean was a very different proposition to a small lake in Ontario, I knew that I felt comfortable on the water. If I'm honest, I'm not sure how I had the confidence to compare a 3,000-mile row across the stormy Atlantic Ocean with a half-mile canoe across a calm freshwater lake – I suspect the boxing match had more of a psychological effect on me than I had given it credit for.

I don't think I really appreciated the enormity of the challenge. Debra had taken one-third of a year to row the Atlantic but even that didn't feel daunting to me. After all, hadn't I already endured a whole year living on an island in *Castaway*? Longevity was not really an issue. And as for the 3,000 miles, that also seemed abstract. I didn't really have a reference point for the distance, the conditions, or the speed of rowing an ocean.

It was, of course, a risk. A huge one. Failure would have been mentally crushing, and potentially career suicide, as I was trying to reinvent myself as a successful adventurer. Everyone thought I was crazy. No one could believe I would really go through with it. But what I lack in confidence and self-esteem, I make up for in resilience and determination. And I have always been drawn to water.

Oceans, lakes, rivers, streams, waterfalls ... I find myself mesmerised by water. I'm not sure if it's the fluidity of the movement, or just the expanse, but I have always found myself calmed and soothed by the water. On it, under it, or in it, it doesn't really matter. I find a spiritual draw towards water and towards oceans in particular. And there is something deeply spiritual and primitive about the ocean. Remember, humans are made up of nearly 70 per cent water. The ocean is part of us. It's no surprise to me that the phases of the Moon can affect our moods. Once you see what it does to the tides, you can imagine the impact on the water within us.

I can still recall every childhood holiday by the sea in St Ives, in Brighton, and on the Isle of Rum in the Hebrides. The ocean was like a friend. Even in stormy weather the ocean never looked threatening; on the contrary, it almost appeared more exciting.

During my year living as a castaway on Taransay, the ocean was both a prison and an opportunity. Each evening I would sit on the top of one of the little hills and look west as the sun set over the vast, seemingly endless Atlantic Ocean. I was conflicted: I hated the artifice of our isolation and imprisonment but I also loved being a castaway. I would stare at that ocean and dream about the opportunity that lay on the horizon once we were set free from the 'experiment'. And the ocean for me has always signified ultimate freedom.

Like a castle without a drawbridge, the island without a boat had become a prison, but with a boat you could go anywhere. I would stare at that ocean and marvel at the idea I could get all the way to Grandma and Grandpa's little cabin on the lake in Canada, or even into Central London by boat. All from the very waters lapping the shoreline of Taransay. If you think about it, the seas are like a metaphor for life. Imagine all of humanity on a single beach; opportunity is ahead of us and it's up to us to choose the journey. We have the option to remain on the beach, or we can set off across the ocean. We can choose our mode of transport – be it a plane, a cruise ship, or a sailing boat, we can make life as easy or as hard as we wish it to be. We can go with or against the ocean currents, with or against the prevailing winds; we can go with a motor or a sail, or we can go on a rudderless raft and allow the currents to take us where they choose. We can even swim it. You see, life can be as tough as you make it.

Maybe it was that unstoppable draw of the ocean, or perhaps it was the desire to test my resilience, but whatever the reason, the idea of rowing across the ocean never seemed too daunting or scary. It sounded more like opportunity and adventure. But while I have always been connected to the ocean, the rowing part was largely a mystery to me.

* * *

The first time I ever went sailing was awful. I was 14 and my friends, twin brothers Bruce and Russell Price, invited me to sail to France aboard their father's sailing boat. It wasn't flash and it certainly wasn't 'crewed'; in fact we did all the hard work in between vomiting and sipping cups of Bovril.

My memories were of seasickness, of being cold and wet, and yet I went back for more. Like a moth to a flame, I found myself drawn to the discomfort. There was something rather exciting about the mix of hardship and satisfaction. I had long admired the heroics of the early sailors. One of the first stories I heard as a child was that of Sir Robin Knox-Johnston who, during the late 1960s, won the Golden Globe trophy in the race to be the first person to sail single-handed, unassisted and non-stop, round the world. His achievement was made all the more memorable by the fact that he was the only one of nine entrants to complete the race, as he crossed the finish line after a mammoth 312 days at sea.

The draw of the water eventually landed me at the University of Portsmouth where my student flat overlooked the ocean. But just looking at it wasn't enough and I soon found myself enlisting with the Royal Navy as a midshipman in the University Royal Naval Unit. We had our own ship, HMS *Blazer*, which we crewed each weekend for small deployments around the Solent. If there was a single series of experiences that helped build my resolve, it was those stormy, cold weekends at sea.

My time aboard HMS *Blazer* gave me my first real test of sleep deprivation and discomfort. We worked in shifts through the night, and we cooked in the tiny sea-sickness-inducing galley. What's more, the permanent crew of Royal Naval sailors who joined us ensured it was no holiday. I owe my toilet-cleaning expertise and my polishing skills to those four years of scrubbing and buffing (and vomiting) aboard that little ship.

During the holidays we would leave UK waters and head down to Spain and Gibraltar, or up as far as Norway and Sweden. I loved and hated every single minute on that little ship, and when I wasn't aboard HMS *Blazer*, I was sailing at the Joint Services Sailing Centre.

My time in the Royal Naval reserves was like an apprenticeship to the ocean. She had become familiar, like a friend, but rowing 3,000 miles in a small 24-foot boat made of plywood and glue was a very different proposition to sailing aboard a Naval Fast Patrol boat with an engine and a competent crew who knew what they were doing.

Back in 2001, when Debra Veal had been catapulted into the headlines, another young sailor had also become a household name: Ellen MacArthur. She had competed in the Vendée Globe, the race to sail single-handed, non-stop around the world, just like Knox-Johnston had done some thirty years earlier. Like millions of other Brits, I had been moved and uplifted by her fighting spirit in the face of adversity. There was something so beautiful about watching this young woman take on 28,000 miles of ocean. I found it empowering and inspiring and I wanted my own challenge. It wouldn't be on a sailing yacht though. The problem with sailing is that the costs are prohibitive; you could spend millions of pounds to get a boat fit for a circumnavigation of the world. I needed something more attainable and achievable. A rowing boat is a fraction of the cost, though at the sacrifice of security and any semblance of leisure time …

* * *

I signed up for the Atlantic Rowing Race two years ahead of the December 2005 start date, and for the first year I buried my head in the sand and pretended it wasn't really happening. Although the ocean held a comfortable familiarity for me, spending many months at sea in a tiny wooden boat was on an entirely different scale to anything I had done before. When the clock ticked round to a year to go, I still didn't have either a boat or, more importantly, a partner.

I was at a party in Central London when I spotted James Cracknell across the room. I recognised him because he had just won his second gold medal in the Olympic Games.

'Now there's a bloke who can row,' I thought.

If you don't ask you won't get, my grandmother always said. Be brave.

I bounded over with Tigger-like enthusiasm and asked James if he'd like to row across the Atlantic Ocean with me.

'Have you rowed before?' he asked, looking at my physique suspiciously.

'Nope. Never,' I smiled. I should have lied.

He told me to get lost, but I persevered and a few weeks later we discussed the possibility of teaming up.

Have you ever met an Olympian? If not, then let me paint a picture. They really are a unique type of person. Stubborn, determined, focused people who have decided to sacrifice their life to the goal of winning medals. It takes a different kind of competitive confidence to embrace a life of professional sport. Imagine dedicating your life to one discipline: jumping, running, throwing, cycling, rowing. Imagine the dedication and the commitment needed to turn up at the gym 360 days of the year. Those freezing cold mornings. The sacrifice of abstaining from alcohol and parties and socialising. I may be over-egging it slightly, but you can't overestimate an Olympian's commitment to their sport.

And the financial rewards are slim. At best you might win a gold medal and fleeting sporting fame, until the next generation come up through the ranks. The likes of Chris Hoy, Steve Redgrave and Helen Grainger are the exceptions to the rule. They are sportsmen and women who have broken the third wall and entered the world of celebrity.

James Cracknell, even among sporting legends, had a fearsome reputation. He was famed for pushing himself and those around him harder and farther than anyone else. He was ferocious. He wanted to be the best. For James it was gold or nothing. There was no point in turning up if he was only going to come second. For 18 years of his life, James had sacrificed a normal life for rowing.

And then there was me. Mr Jack-of-all-trades-master-of-none. For me, variety has always been the spice of life. Oh, and I hate competition and I can't row.

We teamed up.

James and I wrote a book about that journey (called *The Crossing*) but I want to talk about the experience in the context of what I learned from the sea in those seven weeks.

I'll begin with our preparation, or lack of. On paper, we ticked all the boxes. We passed the necessary exams and became Ocean Yachtmasters. If we compare it to a driving test, we did all the theory but failed to do any of the practical. A rowing machine was as close to a rowing boat as I ever got during our preparation.

In the few months leading up to the race, I was filming in the deserts of Namibia.

The morning of the race was only the third time James and I had ever been in a rowing boat together and the previous times had all been photo opportunities. Before we were allowed to start the race, we had to pass 'scrutineering' and here was where our shortcomings really showed. James was asked to demonstrate how to use the handheld pump to bail water out of our boat. He held it the wrong way round and ended up *filling* the boat with seawater instead. Lynn, the chief scrutineer, went red with rage. We had shown an early red flag and she was about to uncover the full scale of our lack of preparedness.

I have never been good with preparation. I have always been one of those individuals who goes in with hope and a prayer. I cross my fingers and hope I'll get through it. My frequent exam failures were proof that it rarely worked. At that stage in my life, preparation felt like extending the misery. I already knew that rowing the Atlantic wasn't going to be easy or even much fun, so what was the point in another year of 'suffering' in the preparation? How wrong I was.

What followed was a baptism of fire, exaggerated because we hadn't prepared or trained ourselves mentally or physically. Don't get me wrong, I was in good physical shape. I had rowed for an hour or two every day for six long months, but always within the safety of my rowing machine. On my terms.

Now we were stuck on a 24-foot rowing boat, *Spirit of EDF Energy*, facing the pressure of expectation. We rowed a routine of two hours on, two hours off, two hours on, two hours off, twenty-four hours a day, seven days a week for seven weeks. We never slept for more than 40 minutes at a time and never more than three hours in any 24-hour block. Sleep deprivation became our nemesis. Our enemy. Irritability descended into madness. I have never experienced such delirium. We became irrational, emotional and vulnerable at the very time we needed to be clinical, focused and strong.

Rowing 3,000 miles of Atlantic Ocean was one thing, but doing it as part of a race with a competitive edge was another. Competition adds pressure, which changes the dynamic. Ever since I was a young boy, I have hated competition. Part of the attraction of the Atlantic Rowing Race was to confront my fears, and the competitive element was an important aspect to that. But it didn't make life easy, particularly as I was with a fiercely competitive Olympian.

For an Olympian like James Cracknell, the race element was bigger than the challenge. Rowing the Atlantic was incidental. For me, the race was a by-product of the task itself, to row the Atlantic. What you must remember is that James wasn't the only sportsman or woman in the race. There were semi-professional athletes from all around the world alongside everyday folk like myself. It was this heady mix of ambition and ability that made the Atlantic Rowing Race such an exciting proposition.

The first week on the ocean was the hardest as we adapted to our new routine. We were still strangers. In the year's preparation we had never really got to know one another. It soon transpired we had two very different objectives. I was there to complete, while James was in it to win it. I didn't care if it took us a whole year to row across that ocean as long as we did so in the spirit of the race, in good health and we arrived alive. James, on the other hand, didn't mind if we died halfway across as long as we had put in 110 per cent effort.

We fought and we wept. That first week was misery. Utter misery, possibly the most depressed I have ever been. The satnav registered 2,995 miles to go and a speed of half a mile an hour. This was perhaps one of the single biggest learning experiences of my life. Those numbers on that machine looked impossible. It wasn't the ocean around us nor the weather or storms that felt overwhelming and suffocating, but the scale in my head.

For the first time in my life, my brain couldn't cope. The enormity of the task was out of my comfort zone and something I had never experienced before. It was abstract. Invisible. Unimaginable.

It was overwhelming. 'Two thousand, nine hundred and ninety-five miles,' I kept repeating over and over. It was just a number. But the figure was like a gremlin. It kept multiplying. I couldn't break down the size of the whole thing.

What kept us both from turning about and heading back into harbour with our tails between our legs? Pride. We were both haunted by the fear of humiliation on turning around. Neither of us could stomach that prospect. Despite the misery, carrying on felt like the lesser of two evils.

I don't remember much from those first few days. It was like waking up to a version of *Groundhog Day* every two hours. There was no escape. When we weren't enduring the oars in the open water, we were worried about the ticking clock to the next session. And on top of that, there were the eyes. James was watching. I don't know for sure that he was always staring and he probably wasn't, but I so wanted to please him. I wanted to prove I was up to the task. I wanted to impress the Olympian. Through the tinted window of the cabin door, James was invisible from the outside, but I knew that from inside the cabin, he could watch my progress as if from a ship's wheelhouse. I always felt those eyes boring into me as I took my turn to row.

I had gone into my charity boxing bout as the underdog. The only way to go was up. In some ways, I couldn't really lose, because as far as the press was concerned, I'd lost before I'd started. But now, here, with an Olympic rower, we were the boat to beat. By teaming up with an Olympic legend, I had put extra pressure on ourselves to perform and to win.

If we failed or lost, I wasn't just letting myself down but James Cracknell too. The shame would be too much. And I found myself obsessing about pleasing James more than anything.

Rowing across the Atlantic was like the ultimate apprenticeship in stoicism. There was not a single moment or break from the relentlessness. We were always tired, or nervous, or sick, or hungry, or irritable, or sleep deprived, or sad, or anxious, or thirsty. Or all of the above. If I'm honest, I can barely think of a single moment of contentment in those seven weeks because I was always worried about what might happen next.

We had to poo into a bucket, and we tried to fill the food vacuum created by 12 hours of rowing each day, which equated to about 10,000 calories. It was monotonous and overwhelming.

I have never really suffered from boredom. I'm pretty good in my own company, but for the first time in my life I had been psychologically outmanoeuvred by the ocean. I didn't know what to do with my mind. I wasn't sure where to take it. The loud inner voice kept screaming '3,000 miles at half a mile an hour!' It was deafening. I had to calm myself down, for I could feel panic and anxiety rising within. I have never battled my own inner voice of doubt as much as I did during those first seven days.

I know it probably sounds glib to say this, but I reckon it would be a breeze if I were ever to do it again (which I won't). It was awful, but the awfulness was largely down to my lack of mental strength. I don't want to destroy the myth, but rowing the Atlantic isn't particularly physical. It is so much more a mental battle of mind over matter, of mind over muscle.

Although I had tested my mental fortitude before, the Atlantic Ocean was on a whole new scale. There was not a single easy day. Life was perpetual uncertainty. Whenever things seemed to be going well, something happened to bring us back to reality.

First the electrics failed, then the water desalinator, which converts saltwater into freshwater, died on us, and then the rollers on our seats broke. The disasters continued. The water-maker broke, the rowing seat broke, our hands broke (with blisters), we ran out of food, the satellite phone wouldn't work. Anything that could go wrong did go wrong. It felt like we were constantly fighting fires. Our emotions peaked and troughed more than the ocean waves.

It was like a never-ending nightmare. Every time we made progress and fixed something, another problem came our way. The breakages and the boat failures began to inhabit my subconscious. I would daydream about things breaking, the loud inner voice of doubt forever looming large.

The result was that almost every action filled me with dread and fear. I can still recall the sinking feeling every time I started the stove, to heat water for our rehydrated meals. If the stove failed or broke, which statistically was very likely in the damp salty air, then we would be stuck. There was no replacement and we couldn't call in a mechanic.

That fear of things breaking made me feel sick. Physically sick. We are so used to calling in other people to help when things go wrong, from the plumber and the electrician to the police officer or the ambulance. On land, we become used to a life of deferring to others, but out there on the ocean we were alone. Really, truly, alone. We stood or fell by our own mistakes. There could be no one to blame but ourselves.

For both James and myself, this was a departure from our normal lives. James had spent his career surrounded by experts, from dietitians to sport scientists. All he had to do was train the muscle and do his thing. (That makes it sound easy, which it unequivocally isn't, but I hope you'll get my point.) I am often surrounded during filming assignments by teams of people who make sure there is food, fuel and water, but here, in the middle of the Atlantic Ocean thousands of miles from land, we were on our own.

The thing about the ocean is that it begins to wear and tear both man and material. Both we and our boat began to fall apart. As well as the blisters on our hands, boils were forming on our bottoms. I never knew how painful boils could be until I had to row 12 hours a day on them. I would fixate about them. But it turned out that the physical and mechanical problems were the least of our worries.

* * *

A near-death experience changed the whole narrative. We were less than 1,000 miles from the finish line when we were capsized by an enormous wave. I was rowing at the time and James was in the cabin. All I remember was my head breaking the surface of the water and the sight of our upturned boat and mountainous seas all around. Strangely, I didn't feel fear. Just resignation. I couldn't see how we were going to get out of this one. So I resolved to meet my maker.

What does it feel like to stare death in the face? Honestly? It feels a bit shit. I was embarrassed at my failing. Typical, I thought. Trust me to fail at life and die too young. There were no angels in the sky like in the 1980s film *Cocoon*. I didn't start to chat to God or a dog. There was no tunnel of light. It was just me and the ocean and a feeling of impending doom.

My past didn't flash before my eyes and I didn't scream out that I wasn't yet done with life. Funnily, I remember looking at the hull of the boat and wondering why there were so many barnacles clinging to her.

James was obsessed about barnacles slowing us down and had insisted that one of us jump into the water each day to scrape the shells from the bottom of the boat. It was unpleasant and uncomfortable. We had argued about it and now I could see that it hadn't even made a difference – the hull of the boat was covered in barnacles, not that it mattered now. We had bigger things to worry about. Like survival.

I am not scared of death. I don't want that to sound blasé, but it's true. I am fearful for my loved ones who would endure the loss, but I'm not fearful of death myself. We have a strange relationship with death. We rarely talk about it even though it is an inevitability for us all. Mortality just isn't very sexy. Death is still the stuff of hushed tones. Setting aside any specifics of spirituality, religion, beliefs or non-beliefs, it is merely the place we will end our lives.

I think we should talk about it more.

In the wilderness, death is a part of life. The line between life and death is much more visceral and tenuous. Nature is red in tooth and claw and the circle of life incorporates death on a daily basis.

In much of modern life, death is an inconvenience. It obstructs the anti-ageing, immortal promise of consumerism. Death doesn't really sell. Immortality, and the market for pills, potions, vitamins, supplements and serums, sells.

Death isn't part of our life script. Until you stare it in the face in the middle of the Atlantic.

I wasn't ready to die. I had too much to live for and I wasn't ready to fail at life itself. Survival instinct is an extraordinary thing. In times of trauma, it really does take over.

So I banished death and opted for life.

I don't recall swimming to the upturned hull of our boat. I can remember hauling myself out of the water and the boat somehow righting itself. I recall the crushing relief of hearing James's voice from the cabin. He was alive. I was alive. But we were drifting in our little rowing boat, alone in the Atlantic Ocean.

We had lost almost everything. Our water desalinator, our solar panels, the batteries and all electrics. We had no GPS for guidance. No satellite phone. No VHF radio for communication. We were lost at sea with little hope of being found.

I was ready to activate our EPIRB, the emergency beacon used to summon rescue, but James had other thoughts. He threw a wet towel at me and told me I had five minutes to pull myself together and get back on the oars before we lost a place in the race.

Despite our near-death experience, James was still focused on the race.

We still had the basics to carry on, including two oars, some paper charts for navigation, a magnetic compass and emergency bottled water in the ballast of the boat, packed for such an eventuality and emergency.

In that moment, my inner voice of doubt disappeared, replaced by certainty and hope. If we could overcome the ultimate ocean test of a capsize, then it felt like nothing was going to stop us.

Interestingly, my surge of confidence coincided with a steep decline in James's physical and mental well-being. He had been plagued by ill-health and, in the absence of a doctor onboard, he had overprescribed himself Tramadol, a highly addictive and hallucinogenic painkiller. With weeping ulcers on his bottom, he was finding it difficult to concentrate, let alone row. James's decline and my strengthened resolve – which came first? I'm not sure if my strength came from his weakness or whether I was just rising to the challenge.

I have often experienced the same dynamic with teammates, almost like a tag team for good days and bad days. I think it is a subliminal survival instinct that keeps you from the double dip, which is when both teammates' emotions dip at the same time and a feeling of despondency and helplessness takes over. That definitely kicked in when we were facing our harshest challenges in the Atlantic Rowing Race, and played a huge part in getting us through to the end.

* * *

After six weeks at sea, I began to look forward to those two-hour sessions on the oars. I could smell and taste the finish. Quite literally. You see, I worked out ways to trick my brain. I trained myself to break the whole journey up into tiny 20-minute goals that would each be rewarded with a piece of chocolate or a fragrant sweet.

I learned to relive experiences in my mind. Like turning on a projector, I would revisit my childhood. It brought light to the darkness and it lifted my morale.

The sleep deprivation was a monster. For the first time in my life I suffered from chronic insomnia. My tiredness began to spiral out of control. But what both James and I learned during that row across the Atlantic Ocean was that there is always more. However low you feel, however wretched the situation, however monotonous something is, the human spirit can always absorb more.

Rowing the Atlantic Ocean was a war of attrition. The sun, the saltwater, the storms, the isolation, the hunger and the sleep deprivation all ground me down. They smothered me, but my spirit, our spirit, was stronger.

If I have one regret, it is that I didn't enjoy the experience more. It felt more like an endurance. I think if I had relaxed more and worried less, then not only would it have been more enjoyable but also a hell of a lot easier.

There were of course moments of great beauty, like the time we were joined by a whale, immense beneath our boat. But they were fleeting. I had a lot of time to think during those lonely seven weeks at sea. It was a frozen moment in my life when I could abstain from outside interference. It allowed for great introspection and moments of clarity.

It was during one of those long nights that I decided to propose to my then-girlfriend, Marina. We had only known one another for a year, but the time at sea made me realise she was the one. I proposed the day after reaching the finish line, with a ring made from a small piece of string from the boat. It was a fitting conclusion to the hardest adventure of my life.

Ten days after the capsize and 49 days after starting the race, we crossed the finish line. The Atlantic Ocean had tested our resolve and we had passed the examination. It was pretty overwhelming. I wish I could bottle that feeling of euphoria when we crossed that line. It makes me a little teary to think back to that extraordinary moment.

Imagine spending nearly two months on edge, surrounded by a deafening noise amid chaos and disruption, and then everything is silent? For the first time in months I could hear birdsong and the sound of waves lapping on the beach. There was no movement. No wind. No fear. No anguish.

It was the greatest achievement of my life. I had never done anything that had stretched me as much as rowing the Atlantic. Fourteen years later, I can still remember the night we arrived at English Harbour, Antigua, at the end of our mammoth journey. The stunned silence as we pulled into our safe haven. The small gathering of family and loved ones. The camera flashes. The tears. The hugs. The sweat and salt. The inability to walk properly on our painfully thin legs.

It is difficult to explain the feeling, but I shall try. Imagine a combination of relief and exhaustion, happiness and elation, numbness and disbelief. It is a pretty heady mix and one that I don't think I will ever experience again.

I really don't think I could ever recreate that feeling of sheer happiness in those few minutes of reunion on the shores of that Caribbean harbour. Boredom and misery and the fear of death were all banished and replaced by safety and security, family and love, satisfaction and familiarity and certainty. The monotonous hum that had been a constant companion over the past seven weeks had been subsumed in a calming stillness.

In actual fact there were quite a few people there to greet us in the middle of the night, but for some reason I remember so clearly the quietness and the whispered tones.

We had finished the race in third place – we were first in the pairs class but third overall behind two crews of four, out of a total of twenty-six starting boats – but the bare statistics didn't really register with me, compared to what I had discovered about myself.

The Atlantic really was a game changer for me. It provided enduring life lessons. I learned to respect nature and to believe in myself. So many of us take our sanitised, clinical environment for granted, but out there on the water we were forced to work with the ocean and not against her. Most importantly, I came to realise that nothing is impossible as long as you believe in yourself.

Lose control of that belief and you've already failed. The Atlantic can be surprisingly forgiving but it doesn't suffer fools. Those seven weeks at sea taught me a huge amount about myself but also a great deal about the ocean herself.

I can't say with hand on my heart that it was an 'enjoyable' experience. There was little obvious happiness or laughter during those challenging days at sea, but I think that was more self-imposed than something the ocean imposed on us.

Above all, I learned the art of self-preservation. Half the battle was self-guarding my own mental and physical well-being. I didn't always succeed, and I battled more demons than on almost any other expedition, but I fought for my spirit and, ultimately, I won.

What is it like living in your own mind for seven weeks without distraction? Once you surrender to the boredom and the notion that you are in it for the long term, it becomes almost cathartic.

That adventure taught me how to cope under duress and how to carry on when all around you in your world seems to be crumbling. I learned to shield myself from the dark cloud of doubt that followed me and James all the way across the ocean.

The ocean can be a cruel mistress. She has claimed many lives over the years. Those 49 days at sea had taught me so much about the nuances of life, the slow wear and tear of mind and body. Like a beautiful dance we had learned to move together with the sea. The ocean had given me the strength to be brave in my moment of fear. We had been lucky. We scraped through despite our early lack of preparation, because the extended nature of the race had given us a chance to catch up as we rowed.

As we adapted to the ocean, as we moved and worked together and with it, I replaced my fear and withered self-confidence with hope, respect and durability. Pitting myself against the ocean gave me the opportunity to blossom into a fully paid-up apprentice who had earned his first stripes in the school of wilderness. I discovered that we are all capable of so much more than we give ourselves credit for. In society we are often held back by presumptions and assumptions. We fail to push ourselves to the limit of our abilities for fear of failure, public ridicule or humiliation. But in the face of the immensity of nature, these concerns begin to pale in significance.

Out in the Atlantic, thousands of miles from society, we were free to succeed or fail on our own terms. The difference being that failure on the ocean could result in more than just a battered ego. Rowing the Atlantic strengthened my self-confidence and taught me to believe in myself. It instilled in me the perseverance to carry on against all the odds. It showed me that I can achieve my goals, even when the going isn't easy. And it demonstrated that if we believe in what we are doing, we can endure whatever comes our way. The ocean had taught me to take control of my own narrative and believe in myself.

CHAPTER TWO

THE ICE CAP

'I didn't know where I was going until I got there.'
Cheryl Strayed

Loneliness, solitude and isolation.

The three words are often used together but they are not mutually exclusive. Most of us have known loneliness at some point in our lives. Some of us have experienced solitude; and during the many months of Covid-19 we have all had a taste of isolation. Most experiences and encounters with the wilderness present themselves with an element of 'aloneness'.

Some people are very good in their own company, while others are afraid of it. Although solitude and isolation can exacerbate the feeling of loneliness, you can also experience them without feeling alone. In some ways loneliness is all in the mind. There are plenty of people who will experience loneliness in the midst of humanity and community. Indeed, there is a loneliness epidemic in the UK in which three million people live alone with loneliness.

I have experienced loneliness, in one form or another, a number of times over the years. There was the loneliness of boarding school which was homesickness. Despite friends and well-meaning teachers, I felt lonely because I was missing the familiarity and safety of my family. While I quite enjoy being alone, I don't particularly enjoy being lonely.

There was the solitude of a year on the island of Taransay during *Castaway* where I was rarely lonely because there were always other people. And then there was the loneliness of a 3,000-mile boat trip down the Amazon where I was surrounded by hundreds of people but isolated by the barrier of communication, or inability to communicate.

To be alone forces you to be more retrospective. Without distraction, you have to face up to whatever the mind fills itself with. This can include fear or hope, happiness or anxiety, doubt or certainty. Without another voice for validation or verification it can be difficult to distinguish between right and wrong.

Isolation may bring doubt but it can also bring the clarity of undiluted, unaffected thinking. The brain and our actions respond according to life-or-death decision-making rather than being swayed by society's moral compass. The result is a purity of thinking that many of us rarely experience. It's like all the background noise is turned off, and the tinnitus of life is replaced by the mellifluous sound of nature.

There are plenty of examples of folk who have gone 'crazy' or 'mad' on their own in the wilderness. Tom Hanks played it to

Hollywood effect in the film *Cast Away* where he makes a volleyball into a 'friend' for company.

I have often found myself having conversations with imaginary people while in the wilderness, talking to myself and even putting on accents of other people. I think these are all coping mechanisms for solitude and isolation. But what can we take away from real isolation and solitude?

Antarctica is the loneliest place I have ever been.

In 2009 I teamed up again with my friend James Cracknell in a race to complete the 500-mile journey across Antarctica to the South Pole. Billed as the toughest endurance race of its kind on the planet, it would be the first footrace to the South Pole since the unintentional one between Captain Robert Scott and Roald Amundsen nearly a century earlier. This time we hoped that the losers would *not* end up dying, as Scott and the remaining members of his team did when they perished in his tent.

I have always admired Captain Scott and his heroic failure. Indeed, there is a history of British explorers and adventurers failing bravely: aside from Scott who died on his return from the South Pole, there was Ernest Shackleton whose ship got trapped in the ice, George Mallory who died on Everest, and Percy Fawcett who disappeared in Brazil.

The notion of racing across Antarctica was as alien as it was scary. I had read endless stories about Antarctica, all of which involved storms, gales, hypothermia, starvation and frostbite. Antarctica was the testing ground for real explorers, and while I wasn't a 'real' explorer as such, I had proven myself on the

Atlantic and I was feeling bold enough to take on the South Pole. The race would involve teams from all round the world. James and I needed to find a third teammate. We decided to put an advert in the newspapers to find a companion mad enough to join us.

I suspect that the subliminal idea of the advert came after the famous one placed by Shackleton seeking to recruit for his 1914 Antarctic expedition, which read: 'Men wanted for hazardous journey. Small wages, bitter cold, long months of complete darkness, constant danger, safe return doubtful. Honour and recognition in case of success.' I bloody love that. There have been some doubts expressed about its authenticity but it always lifts my spirits nonetheless.

Thousands applied to our modern-day advert and we eventually settled on Dr Ed Coats. James and I were thrilled that we would be accompanied by a doctor, until we discovered that Ed was a gynaecologist! By then we were committed, but it didn't matter. Ed was a great teammate.

The race would take us 500 miles across the barren wastes of Antarctica to the South Pole. Antarctica, the seventh continent, is one of the most hostile, cold, windy, arid and highest places on earth. It is an uncompromising, do-or-die environment and arguably the most isolated and remote place on this planet. Apart from a handful of scientists and researchers who live in research stations, Antarctica is devoid of human life, and this brings with it a powerful sense of isolation.

Our first destination was the Russian base of Novolazarevskaya, or 'Novo' as it is known, located on Queen Maud Land, some 2,500 miles from the closest inhabited township, Tierra del Fuego on South America's southern tip. Total summer population of Novo: 70.

We flew in aboard a huge Russian cargo plane. The loo consisted of a portacabin strapped to the floor. It was as surreal as it was exciting, stepping from that aircraft into the minus 25 degrees Celsius chill, particularly as we had departed the warmth of the South African summer in Cape Town just a few hours earlier.

It wasn't until that plane departed that I got a true sense of what isolation really meant. No further planes would land for six weeks. People have been known to go crazy watching that final plane leave. I felt abandoned and alone. More alone than I have felt almost anywhere else. Despite my teammates, our competitors, organisers and even the film crew that would document the journey, the fact was that we were alone on this vast continent. The umbilical cord with the outside world had been severed. Loneliness is often all in the mind. I didn't go crazy but it certainly had a profound impact on my sense of solitude and isolation.

Once the race began, we quickly split up and we rarely saw other competitors. To all intents and purposes, we were alone. The three of us on a continent of five million square miles.

The descent into solitude and loneliness was gradual. For 18 hours each day, we would march, head down, pulling our sledges loaded with the vital supplies on which our lives relied, into the biting polar wind that sometimes reduced the temperature to minus 50 degrees Celsius.

There were days when the wind and cold created an impenetrable barrier between the three of us. We couldn't talk, as our words were lost in the wind, and we couldn't see one another because of the hoods, goggles and balaclavas to protect us from frostbite and hypothermia. It was on those days that loneliness became both friend and foe.

To cope, I reflected on my previous experiences with loneliness. I had experienced loneliness in a number of forms while rowing the Atlantic and living on Taransay for *Castaway*, but the loneliness had been uniquely different each time.

Castaway had been a loneliness among others. Rowing had been a loneliness of routine, and here, in the greatness of Antarctica, it was a loneliness of isolation and remoteness. My experiences of loneliness had built a resilience. I had learned how to cope with it. It was no longer a fearful sentiment that left me anxious and sapped of ambition.

But while I had become accustomed to the solitude of isolation, Antarctica has a way of weakening and denting that spirit of resilience. The combination of sleep deprivation, hardships and discomfort created a debilitating mix of anxiety.

We were on our own. We had no one to turn to when the going got tough. Although we had trained and prepared for the expedition, the hardships should not be underestimated.

Let me paint a picture. We each pulled a sledge, or 'pulk' as they are known. Those pulks, weighing up to 80 kilograms, contained our lives: a tent, a stove and fuel. Without these, without the heat and shelter this equipment provided, survival was impossible. The fuel in the stove was used to melt the snow and ice for water that was used for the hydration of our bodies and for rehydrating our freeze-dried food. If the fuel or stove failed, we would die. The tent would allow us to sleep, protecting us from the bitter cold wind that ripped across the icy plateau. Although just a thin fabric, it was the difference between life and death.

The tent was also our barrier against the storms, that lashed us with such ferocity that we each took turns to dig ourselves out of the swiftly building snowdrifts. Every half an hour, one of us would need to weather the storm and dig the snow mountains that built up around the tent. If we failed to clear the snow, it would eventually overwhelm us and smother us to death.

And the dangers didn't end there. If we wore too few clothes, we could become hypothermic and die. If we wore too many, we might sweat. The sweat would then freeze and the freezing layer of icy shirt against your skin would bring on hypothermia and you'd die.

If we got lost, we would die. If we stumbled across a crevasse field among the glacial plateau, we risked falling into a chasm that might open a kilometre into the belly of Antarctica … and we would die.

I don't want to sound too sensationalist, but this was the 'theatre' and the backdrop to the loneliness I felt.

Antarctica is a harsh, unforgiving wilderness. Almost barren of life and even of features. There is an exquisite beauty in this nothingness, but it can be a tough place to love.

Most wilderness, however rugged and harsh, has a redeeming feature, perhaps the elevation or the colours or the flora and fauna. The mountains, ocean, jungle, and even the desert have variety, inconsistencies that are interesting. Even the Arctic has the benefit of icebergs and blue ice among the whiteness, but here in the Antarctic polar desert there was just nothingness. The nothingness is, of course, a part of the allure, but it is also the environmental equivalent of being in a sensory deprivation float-tank.

There were days when the whiteness of the snow bleached into the whiteness of the sky. There was no horizon. You couldn't see anything, there was no depth of field. You couldn't even tell which way was up and which way was down. You relied on gravity to keep you on your feet.

Imagine stepping into an oversized snow dome and shaking it up. Just whiteness and nothingness. Now add to this feeling of nothingness hunger, fatigue and sleep deprivation and you have the beginnings of understanding the extreme Antarctic experience.

For me, there was rarely fear (except for the time that James led us into the middle of a crevasse field and we heard the ice bridges beneath our skis crashing into the invisible abyss below), but my increasing sense of isolation led to a growing feeling of loneliness.

By now James had become a mate. A real friend. We still had a complex relationship because of our very different characters, but he was no longer the stranger he had been on the Atlantic. Ed, too, had become a friend. Solid and dependable, he had brought stability to the sometimes fractious relationship between James and me. That said, I sometimes found myself absorbing Ed's struggles with James. I ended up absorbing his own inner battle vicariously and living it myself. It was one of the first times I realised the importance of team morale. If one person is unhappy and miserable, it has the power to 'infect' the whole team. Morale really is contagious. A positive attitude was every bit as important as the rest of the equipment we were carrying. Despite my best attempts to remain a force of positivity, the environmental deprivations had begun to find a weakness in my resilience. Once Antarctica finds that weakness, she exploits it, and eats away at it until what starts as a tiny scratch becomes a large weeping wound.

A big part of the problem was the boredom. I had experienced boredom plenty of times before in my life and had always prided myself in my ability to adapt and occupy my mind, but for some reason, the nothingness and the whiteness and the isolation, both personal and geographical, began to weaken my spirit.

We humans need rewards. A chocolate bar, a sunrise, a kiss, a letter ... we need to have something to look forward to, and here in the midst of a race across Antarctica, those rewards became more and more elusive. There was no music or conversation or podcasts or talking books (the icy cold stole the power from any batteries). Often there was no view, not even a horizon. Even the end of each day promised little reward.

In fact the end-of-day stop would necessitate more effort than the 18 hours on the skis. We would need to erect the tent, shovel snow, take turns melting water, and on top of this three large, smelly, irritable men would need to squash into a tent meant for two. And then all we had was a short sleep from which we would need to start all over again. Indeed, there were times, particularly in high wind, when I actually dreaded the stop.

The result was that time seemed to slow down to a half pace. I could almost hear the second hand ticking in my head, taunting me with its idleness. It reminded me of those first few hours on the Atlantic Ocean. The overwhelming magnitude of the journey ahead, the perpetual, seemingly never-ending challenge. In any other circumstance, I would have welcomed the luxury of time. How often have I wished I could press the pause button on life and have time to breathe and appreciate the moment. But not here. Not in the frozen wasteland of Antarctica.

I worked out ways to control the rising anxiety from within and I began to perfect the art of recall, whereby I would lose myself in happy memories. I often worry that I have filled my life with so much, that I don't have enough time to digest what I've done. I'm so busy filling the 'now' that each experience is in danger of disappearing into the ether. But here in this solitary place, I was able to harness those memories and experiences as a powerful tool to feed my ebbing spirit.

What is this 'spirit'? For me it is the key to almost everything. With spirit comes happiness and hope and love and ambition. Spirit is more than just an internal emotion. For me, spirit is as much a part of your environment as it is a part of how you feel. It is fed by personality and experience and friendship and family and love and death.

Spirit is the invisible us. It is the voice in our head. Our highs and our lows. Our laughter and our sorrow. There are people who are born with 'spirit', and there are those who build that spirit. The power of the environment in the creation of spirit should not be underestimated. The wilderness, through the weather, the flora, the fauna, the solitude, the 'organic-ness' and the peacefulness, has the power to both fuel and heal spirit.

My spirit soars in the wilderness. Uncluttered and undistracted, I feel so light I could fly. I have a firmness of purpose that I rarely have back in city life where, weighed down by the conventions and convictions of society, there is a constant din. It is like a cacophony of negativity. In the wilderness, the comparative silence is simultaneously soothing and enabling.

Here in Antarctica, although I didn't realise it at the time, the experience was nature's way of rebooting my waning spirit. As I battled against the wind, I would revisit my childhood. Take a walk through my childhood home. Visit Dad in the veterinary clinic. I would amble through my old school, say hello to some teachers and meander along the river. I would cuddle my dog and even go on a date with Marina. I would go so far as choosing the table and then the food itself. I could taste it. I could feel Marina's lips on mine. It was extraordinary, the clarity of those hyper-real memories. I had never experienced them before and have never had them since, but at the time in Antarctica it was like divine intervention. All of this against a backdrop of infinite whiteness.

Of course it wasn't always a blank canvas. There were days when the clouds lifted, the wind dropped and the sun cast her rays across the icy plateau, highlighting fifty shades of white. The wind would carve the snow and ice into frozen waves that would take on all sorts of shapes and silhouettes in my hallucinatory sleep-deprived consciousness.

Sometimes I felt drunk, high even. The isolation and blankness of the wilderness gave me the clarity to recapture lost memories. It was like finding a long-lost book or letter and reading it again. Modern society is so obsessed with the next thing that we fail to appreciate the past. So many of us are so busy looking at the horizon that we fail to look back. That's not to say we need to dwell and loiter there for too long, but we must look back to look forward. How can we avoid making mistakes if we don't learn from our history?

Those solitary, lonely weeks in Antarctica gave me time to appreciate how far I had come in life and how lucky I had been. From the discomfort of my environment, I was able to sensibly and rationally digest and appreciate what I had in life, rather than wishing I had more. It is so easy to be blinded by complacency, but here in the nothingness of the polar landscape, the wilderness gave me the luxury of time and the clarity to go on.

Like the flame of our stove, the spirit that has driven me through life began to grow, and with the fire came great warmth.

We didn't win the race to the South Pole. We were beaten by the Norwegians. Those pesky Scandis are pretty good on skis and know a thing or two about polar competitions – after all, it was the Norwegian explorer Roald Amundsen who beat Captain Scott to the South Pole. Nonetheless, we finished the race in second place.

We returned home skinnier and stronger. Antarctica really had toughened me up. It had tested me in ways that the Atlantic hadn't. Isolated from the rest of the world, I had time to contemplate my own life. It had reminded me of the importance of micro-managing my very own existence.

I even managed to save my nose. In the final few days of the race, the cold wind had begun to attack my nose and to freeze the skin. It started as a white, peeling section of skin, which soon began to darken. Without a mirror I was oblivious to it, but it was Ed who first noticed it. Now the idea of losing part of my nose wasn't something I took lightly. I began to obsess. Fortunately, I managed to work out a practical solution to the problem and used my extra hand-warmers, which I strapped across my face with a bandage, to soothe the affected area. As soon as one hand-warmer began to cool, I would replace it with another. I micro-managed the problem to avoid frostbite and somehow saved my nose. I still have a small scar as a reminder of the importance of self-preservation and taking responsibility for your own body.

We might not have won the race but I had fallen in love with Antarctica. I was certain that I would never return, but a few years later, in 2011, I was invited to go back to make a documentary for the BBC about Captain Scott. This time I was a guest at New Zealand's Scott Base, from which we travelled overland to Scott's hut.

More than a hundred years ago, Captain Scott and his team had sailed South to Antarctica where they had erected a prefabricated rectangular hut, 50 feet long and 25 feet wide. It was built at Cape Evans on Ross Island and it would be their home for nearly two years, their base camp while they planned their expedition across the ice to the South Pole. They used seaweed to insulate the cabin and acetylene gas for lighting.

After Scott's death in 1912, the hut remained untouched until it was dug out of the snow and ice by a US expedition in 1956. They found the hut almost entirely intact, exactly as it had been left, frozen in time, apart from the rising damp from the warming temperatures melting the ice. I was there to make a film about the conservation work of a team of New Zealand conservators working to protect the cabin and its precious contents, and I would join them and use the hut as a way of re-telling the tale of heroism and disaster of Scott's final mission.

I shall never forget stepping into that hut. It was like a living museum.

The hut itself had been weathered a light grey by a hundred years of Antarctic storms and 24-hour summer sunlight. It was amazing to me that it had remained intact for so long in this harsh environment. Perched on the shore of a frozen ocean, it was the only sign of civilisation in this bitterly cold and inhospitable place.

I have always loved cabins. I find there is great romance to a wooden structure built seamlessly into nature – the manifestation of a more sympathetic relationship between man and his surroundings. As I squeezed the wooden latch and pushed the door open, I was overwhelmed by a musty smell. It was unlike anything I had ever encountered before: a mix of wood, smoke and decay. I would soon learn that it was created by a cocktail of old blubber and horse hair.

My eyes struggled to adapt to the low light. It was like stepping into a time capsule: there were packing crates, food, jumpers, beds, sleeping bags, skis, sledges, jars, bottles, photographs, cameras, bits and pieces of scientific equipment. Imagine leaving your house now, in this instant, and then returning a hundred years later. That was how the hut had been left, and it was as overwhelming as it was moving.

The hut had been divided into two sections. A wall of packing crates had been erected between the sleeping area and the living area. There was an outdoor area for washing as well as stables for the ponies and kennels for the dogs. While I too had endured the hardships of Antarctica, the hut was a reminder of the sacrifices made by early pioneers to help us understand the wilderness. There was a sadness in that hut. I'm not sure if it was magnified by my own sentiments, or my understanding of its historical significance, but I felt it. The isolation and loneliness of Antarctica are palpable. They are always there, and perhaps knowing the tragic ending of those who spent years living in that hut added to the burden I carried within. I do know that during the shooting of the documentary. It always felt happier outside the hut than inside it.

The history of British polar exploration is dominated by two great heroes, Robert Falcon Scott and Ernest Shackleton. For those who have studied or read about the two great titans of exploration, you tended to be either in the Shackleton camp or in Scott's.

Scott, the Royal Naval officer, was a man of discipline. A hard man, he stuck to his decisions and maintained a distance between officer and man. Shackleton was a Merchant Navy officer. Softer, gentler, his style of leadership was one of equality rather than division.

Their differing styles are evident in their own unique stories. Shackleton, the man whose ship got stuck, crushed and sunk in the ice. He escaped with his men to Elephant Island where he then set sail in a tiny open boat to South Georgia for one of the greatest acts of daring in the history of exploration. He made it, and from there he was able to organise a rescue for his men. Scott, the man who made it to the South Pole only to discover he had been beaten by the Norwegian Amundsen, and who would subsequently die on his return journey.

Not far from Scott's hut is Shackleton's at Cape Royds. A much smaller shelter, it too is frozen in time and a lasting legacy to an extraordinary explorer.

What surprised me most was the different manifestations of spirit and hope within each of the two huts. Whereas Scott's was dark with a heavy sadness, Shackleton's was light and airy, there was no division between teammates and the hut held a positivity that had been lacking in Scott's.

What has this got to do with the wilderness, you might ask. Well, I think it is important to remember that much of what we experience is from the baggage we carry. Were those sentiments really coming from the huts themselves or from my own experiences? Not only did I know the stories behind the huts but I too had suffered and endured the weather outside the hut. I had experienced at first hand the violence that Antarctica and the polar regions can wreak. I could imagine the fear and the trepidation of those great explorers.

A century ago, none of those men really knew what lay beyond. They had no idea what they would have to endure. The South Pole was still a large unknown, and they were the pioneers opening up a new world, heading into unchartered territory.

Many people escape to the wilderness, but you can never escape yourself. We have internal baggage that dilutes and tarnishes how we experience things. The feeling of isolation and solitude is often more internal than external. We talk of people who are happy in their own company. There are some folk who thrive in isolation while there are others who feel burdened by the notion of solitude.

The huts of Scott and Shackleton gave me the chance to explore my own inner manifestations. I found myself wandering around these living museums, absorbing the spirit of history, hope, optimism and despair. The huts were like a tiny portal into a simpler, humbler way of life. Not that either man had been slumming it. Both cabins were stocked with whisky, champagne, beer and wine. There were cases of duck liver foie gras and tins of baked beans and hoosh (a meat stew used by explorers of the time). While we had lived off rehydrated packs of food during our race to the South Pole, the early polar explorers had ensured that they thrived, not just survived.

There is a notion that the wilderness is always a battleground. But this couldn't be further from the reality. That month I spent back in Antarctica was one of the happiest, most beautiful experiences of my life, and it gave me the chance to experience another face of the South Pole.

While my first visit had been one of endurance and suffering where a handful of nuts had been my only escape from the monotony and hardships, my return visit had been more controlled. Armed with bottles of whisky and access to the internet, the feeling of isolation, or the perception of isolation, with a tiny team of people again on the same remote, vast continent, was entirely different. Every morning I walked along the frozen shore, past the beached icebergs soaring into the pristine blue sky, past penguin colonies and across meltwater streams, struck by the beauty of this other-worldly landscape.

When I think back on all the things I have done, that month at Scott's hut remains one of the happiest, most uplifting of all my wilderness experiences.

* * *

Cabin huts in general are often synonymous with solitude and isolation. There is a simplicity to cabin life that is deeply attractive. While Scott and Shackleton's cabins took on the attributes of Antarctica in their very remoteness, there is a spirit of isolation and 'apartness' that comes from cabin life in general. Over the years, I have spent long periods of time living with families who have retreated into their own cabins. They have adopted a new, slower, more organic pace of life, and there is something hugely appealing about that simplicity.

As a child I spent my summers at a wooden cottage hand-built by my late grandfather on the shores of a lake in Eastern Canada. In fact, it wasn't a cabin but it was a cabin-like structure. It was simple, wooden and wonky. The floorboards squeaked, it smelled musky, it was uneven. And it was glorious.

Go onto Instagram and you'll find hundreds of accounts (with millions of likes) dedicated to cabin life. There are countless coffee-table books that can be displayed in million-pound townhouses, as an ode to a simpler life. There are of course countless international cultural alternatives: the ger or the yurt of Mongolia, the tepee of North America, the adobe roundhouses of Botswana and the stilted longhouses of Borneo. The cabin is as close as we get to living in harmony with the environment. It has a sense of belonging, and with that belonging comes happiness and simplicity.

Over the years I have been lucky enough to visit numerous cabins around the world. But now, as these words flood the page, I can picture myself back in a cabin in the Arctic Circle. I am with my family. They are next to me now. My son, Ludo, is reading his book under candlelight while Marina and our daughter, Iona, are playing cards. It is 2 o'clock in the afternoon and it's already pitch black. It is mid-winter, and at this time of year the sun hardly ever rises here. For half an hour each day, the landscape is bathed in an extraordinary pink light, but for the other twenty-three-and-a-half hours, it is as black as the darkest night.

Our cabin has no electricity, no running water and no central heating. Candles give us our light, while a wood-burning stove gives us heat and melts snow and ice for water. We are many miles from the nearest community in a remote corner of Sweden. On the face of it, we should be bored rigid, at each other's throats, wracked by cabin fever and depressed by seasonal affective disorder from lack of light.

But we are not. We are happy. Deliriously happy. Heavenly happy. So happy.

It is as if the noise of the world has been turned down. Not just the silence of the forest in which we are living, its noises absorbed by the snow that blankets the landscape, but the silence of isolation. Beyond the tentacles of the grid, we are alone. Really alone. There are no distractions to drive wedges between us. We are together at peace with ourselves.

In some ways, cabin life is like a little gateway into a parallel universe. Not far from us, just a few hours along snowy roads, is Swedish civilisation. Cafés and lattes and Ikea and meatballs. But here we are stripped to our basics. We have companionship. We have the wilderness. We have solitude.

Our day is broken into a simple routine. I rise at 4 am to light the wood-burning stove to warm the single-room cabin that has chilled to minus 5 degrees C overnight. Outside it is minus 25. I fuel the fire each hour until my family begin to rise after 9 am.

There are no alarm clocks. No timetables. No work or play dates to race to. It is like a pressure value has been released. I don't find myself disappearing into the wormhole of social media. There is no distraction from the radio or angry news headlines.

We eat breakfast, fried eggs cooked on the stove and fresh coffee, and then we settle into our books. At midday we venture into nature's freezer outside to enjoy the fleeting flash of winter daylight. As we trudge through the snow, we feel like the only people in the world. I have rarely felt such peace and happiness as I feel now.

As the cold begins to bite into our feet and our fingers, we return to the warmth of the cabin. Out here in the wilderness, the simplicity of the cabin becomes a sophistication. She is our sanctuary from the bitter cold. She represents warmth, safety and comfort.

The cabin is not mine, she has been lent to me by one of my favourite people on earth, Olly Williams. Olly is an artist. A bloody brilliant artist, but he is also one of life's happy people and that happiness is etched into every fibre of this place, which was built for his family. I get a little teary-eyed when I think of how much love there is in this cabin. At this moment in time, I never want to leave.

For many of us, happiness is retreating into the innocence of our childhood, to a simpler time free of complications. It comes back to the question of what happiness is. Is it the journey or the destination? Is it something we spend our lives working towards, or is it something that is everywhere but nowhere?

As we grow older, like a gnarled tree, we grow more cynical and wary. We have been stung by the realities of life, and with that cynicism comes a heaviness that often masks happiness.

When I cast my mind back to those long summers in Canada, I can visualise the cabin next to the lake. It is of course always sunny (which it rarely was; it always rained, but the mind has a beautiful way of forgetting the bad times and remembering the happier ones). The sun reflects off the fresh water of that lake, tantalising me to come in. I can see a young me squealing with delight as I leap from the wooden jetty into the sparkling water. The bracing chill of that fresh water on my skin. I can see myself paddling Grandpa's red cedar canoe. Past pine trees whose errant branches dip into the water. I can feel the dew from the grass beneath my bare feet as I race around the cottage, building a den or exploring along the shoreline. Why is it that those memories are so vivid and happy? Visceral, almost. I can feel, smell and taste it. And why does it remind me of so much happiness? There must have been boredom and arguments, but I don't recall them. Perhaps it is rose-tinted glasses as our brains search for happiness.

If those childhood memories are so powerful, perhaps it is no surprise that so many people spend their lives trying to recapture the happiness of their childhood by regressing into their own simple worlds. The wilderness, for many, is a tiny window into those memories of happiness.

It was Shackleton who once said, 'By endurance, we conquer.' The memorial to Scott in Antarctica is a quote from Alfred Tennyson's poem, 'Ulysses': 'To strive, to seek, to find and not to yield.' I find both of these deeply moving.

To endure is to become someone. To endure solitude and isolation builds character and strength and patience, and with patience comes the ability to endure. It's self-fulfilling. Where there is no struggle, there is no strength. When you break life down to its bare basics, you strip out your own expectations with it. Back in society it is easy to get bogged down in the minutiae of your own life and other people's lives. How often do we find ourselves absorbed and drawn into tittle-tattle gossip about others? It's like a vacuum we need to fill for comfort, but stripped of that comfort we become more primal, we are forced to think about the raw basics of life, and the struggle to live day to day becomes all-consuming.

Life in the wilderness is all about the here and now. There is no time for idle thoughts about the Beckhams and the Kardashians, or about the latest Donald Trump tweet. Being absorbed by the very simple equation of life or death, strips us of the need for shallow, unnecessary information. Maybe it's why I always feel cleansed after expeditions. It's like the wilderness has purged the mind and body of anything unnecessary, together with the anxiety that goes with it.

The wilderness reminds us to focus on the essence of our lives rather than comparing ours to other people's. If you think about it, so much of the daily information we absorb is about other people. We are constantly forced to compare and contrast. I suppose it is a good way to keep us in check, to nurture ambition and make money, but it isn't healthy for anyone. Comparing your own life to others is the death of joy and happiness. It's like an act of violence against yourself.

There is no place for comparison in the wilderness. Flowers

and trees don't compare, they just flower, blossom and bloom. That doesn't mean there isn't competition for water and light, but break these down and you are talking about life and death. And ultimately, that is what it is all about. Survival. We have lost a sense of perspective. Society dwells and obsesses about the comparatively unimportant, whereas an expedition or time in the wilderness forces you to focus on what's happening right now.

The wilderness is like a brake. It slows us down and reminds us what is really important in our existence. Society teaches us that life is a race, a competition against others and a battle of comparisons. Whereas the wilderness reminds us that life is a marathon, not a sprint, and that life is there to complete, not to compete.

CHAPTER THREE

RIVERS

'When one tugs at a single thing in nature,
he finds it attached to the rest of the world.'
John Muir

A river cuts through rock not because of its power, but because of its persistence.

I have a vivid childhood memory from when I was about 11 years old. I am alone on the banks of a small river, the Arun, in West Sussex. I can hear the birdsong and the hum of insects. I can feel the summer sun warming my skin as I stare at the slow-moving brown water. This is one of my earliest memories of liberty and freedom, allowed by my parents to walk across the field from our small farmhouse to spend the day fishing alone. I would stare at the water, occasionally throwing in a small stick to watch it being swept gently downriver. There was a small bridge from which I would play pooh sticks, where you'd drop a stick on one side of the bridge and watch it as it got carried downstream with the movement of the water.

I can still taste and smell that idyllic moment in time. I recall vividly the hypnotic flow of the water and the excitement of occasional unidentified splashes. I can still feel that tinge of excitement at what might have lain beneath the surface – the mystery of the underwater world as mesmerising as the movement of the water itself.

There is something magical about a flowing river. It is reassuring and timeless. A river has a consistency that is often lacking in other watery spaces. Of course it can change in level and even shape, but its consistency is something I have always found comforting.

As a child, I loved fishing. I could spend hours on the river bank. I don't ever recall catching a fish, but that was never really the point. It was the ceremony and the thrill of the unexpected happening that would keep me there, alone, for hours and hours.

When I went away to boarding school in Dorset, I cried with homesickness for a year. The only consolation was the River Stour that flowed through the school grounds. I would spend hours sitting on the river bank. That river was my connection to home. I used to imagine how I could climb into a little boat or canoe and paddle all the way down the river to the English Channel, head east along the coast and then up the River Arun to Mum and Dad's house.

Even in London I used to walk along the Thames and marvel at the idea that I could reach my beloved Canadian grandparents in the heart of Canada from Central London. I would draw a line with my finger on a map of the world, down the Thames to the English Channel, down the Solent, across the Atlantic and into the St Lawrence Seaway, down through Lake Ontario and then up the river to Peterborough and into the Kawartha Lakes and finally to our cottage on Lake Chemong. Rivers were like the arteries that bound us together. To this day, rivers have always been a link to friends and family.

There is also the allure of their one-way system. Rivers never go in reverse, they don't live their life in the past. They only look forward. I sometimes think of rivers as a metaphor for life. You can swim with the current or against it. It is almost as if the world is divided into those that take the safe certainty of going downriver with the current, and those who battle upstream against it.

Then there's the silence. A river can be deep and mighty and full of life, and yet there is also peace and tranquillity in a meandering river that belies its power and strength. And there is the fascination of what lies beneath. As a young child, I was a bit obsessed with the paranormal, and with sea and river monsters in particular. Deep down I knew there was no such thing as Nessie, the Loch Ness monster, but the dream was just too good to abandon and I often convinced myself of alien species living beneath the murky brown waters of the Arun.

Little did I realise then, that thirty years later I really would be swimming with monsters – but more about that later.

* * *

When I was younger, I never asked myself what it was about rivers and streams and canals that was so captivating, but I have often wondered since.

Of course, there are lakes, lagoons and ponds, and oceans and seas, but rivers have an attainability. The line between water and land, unpredictability and certainty, danger and safety, is visceral and literal. Rivers are organic, living roads transporting their inhabitants from one place to another. I'm sure it's one of the reasons for my enduring love of pooh sticks. There is a perpetual movement that signifies journey and hope. And the notion of a beginning and an end is entrancing: where does a river start and how? How does it reach such volume? Where does it end?

Of course, the source and the mouth of a river are equally steeped in mystery. Rivers begin on mountains, at the end of glaciers, with simple ground springs. I made a short series for the BBC about the source of the River Thames. We travelled all the way through Gloucestershire to a meadow and a small patch of muddy ground, and that was it, the source of the mighty Thames, around which London has become one of the most important cities in the world.

I found it fascinating that such a mighty river can start from a tiny puddle. Never underestimate the power of growth; like the building blocks of my own character, it gave me great hope.

Then you have the confluence, where the river reaches the sea, or a lake or an ocean. That area of brackish, briny water where the two bodies of water meet and merge fascinates me.

Over the years I have explored plenty of rivers all around the world. Shortly after Marina and I married in 2006, I took responsibility for organising our first summer holiday together as man and wife. I thought a rafting holiday in Sweden would be a great idea. Of course, no Fogle holiday would be complete without a dog, so we took two – my black Labrador Inca and Marina's chocolate Labrador Maggi.

We drove for two days to central Sweden where we were presented with a tonne of timber to construct our raft. For 12 hours we heaved 70-kilo logs into the river and lashed them together with rope, then launched our little raft and set off down the river for a ten-day 'float'. It was ambitious, particularly given my awful track record in knot tying!

To start with, it was idyllic – we launched just as the sun began to lower towards the horizon. Hope and happiness swirled around that little raft as we swept effortlessly past magnificent Scandinavian forest through the chilly waters. But our utopia, a little like my knots, would soon begin to unravel.

The first problem was stopping. We had ambitiously brought our dogs along for the adventure and we soon realised that they would need to be walked and exercised twice a day. Sounds easy? Well, perhaps yes, until you try to stop a rudderless one-tonne raft from being swept downstream by a 3-knot current. Doesn't sound much? Believe me, 3 knots is fast. Paddles were useless in that situation.

Eventually, I tied one end of a rope to the side of the raft and the other around my waist. As we neared a bend in the river, I leapt into the numbing waters and swam hard for the shore, hoping that my calculation was roughly correct. If I could get out of the water, find a sturdy tree and lash the rope to it before the raft rounded the bend, I could haul her to shore.

Breathless from the swim, I raced up the bank and found what I thought was a suitable tree. I secured the rope around her trunk, and stood back. As the rope began to tighten, so I could see the increasing tension in the rope from the force of the weight of the raft as it neared the river bend, and then …

SNAP.

The combination of weight and movement was too much for the rope and tree. I now had to catch up with the raft … and Marina who was now sailing alone with the dogs. Have I mentioned that Marina hated being on the raft alone?

At the next bend in the river I tried again. This time, I buffered the rope a little to absorb the weight. The tree and the rope creaked and groaned but somewhat miraculously the raft came to a halt before swinging around towards the river bank. Success! Except that I needed to do this twice a day.

Then there were the eddies. Oh, how we hated those eddies. They appeared almost from nowhere, as the raft hit a tight bend where the flow of the water was pulled back on itself in a sort of wide, invisible whirlpool that would hold us in its grip. The raft would circle dementedly, caught in the thrashing jaws of nature. We would both take a paddle and sit on one side of the raft, trying to anticipate a possible exit area from the eddy, at which point we would paddle furiously to try and escape her clutches. Many times we would be swept in an unstoppable arc back into the maelstrom. It was as demoralising as it was humiliating, but we would persevere and eventually escape.

As if things weren't bad enough, our adventure also coincided with one of the worst Swedish summers on record. Temperatures rarely broke into double figures and I swear we saw snow. A few days into our adventure it became apparent that my rope-tying skills weren't up to standard. They began to unravel – as did the raft. Soon we were up to our ankles in freezing river water, sheltering under the tiny awning with the dogs.

Cold, tired and irritable, Marina eventually lost it and we stopped the raft. Dressed in a bikini under her raincoat, she marched up to a remote Swedish farmhouse, knocked on the door and begged for shelter. The kind farmer let us sleep in his barn for two days while a record amount of rain fell. We felt like Mary and Joseph sleeping in the hay. He fed us and we eventually made the decision to abandon the raft, leaving a poor Swede to navigate our sorry float to the end of the river where it was disassembled.

I haven't been allowed to organise a summer holiday since. But we learned a lot about rivers. And in the years since, I have explored plenty of rivers around the world, on paddleboards, rowing boats, sailing boats and in canoes. But perhaps the most extraordinary river journey of all was one in the heart of Africa, a journey that would take me beneath the surface and face to face with one of the world's most fearsome predators.

Africa happened by accident. I was in Botswana with Princes William and Harry, following and documenting some of their conservation work. On our final day, we ended up in the Okavango Delta where we were camping in the bush.

As we sat around the campfire, one of the local guides, Brad, told us about the time he ended up swimming with crocodiles. A keen scuba diver from South Africa, Brad had been curious about exploring the mysterious underwater world of the Okavango, but the hippos and the crocodiles had put him off, until one day he decided to take a gamble. He put on his scuba gear, and in he dived. Just a few minutes later and a mighty Nile crocodile swam up to him. He looked at it and it looked at him.

He was pretty sure that he was about to become dinner, but nothing happened. Slowly he edged away and made his way to the surface. He climbed out, took off his mask and breathed a huge sigh of relief.

Why hadn't he been attacked? Was it luck that he hadn't been eaten? Maybe the crocodile just wasn't hungry? The only way to find out, he somewhat bravely decided, was to get back in the water, which he duly did, and once again, a crocodile approached. There was a stand-off but the crocodile remained poker-faced, uninterested.

He couldn't be sure, he explained, but he had hypothesised that the reason the crocodile had left him alone was because it didn't know what or who he was. It didn't recognise him as either predator or prey. He wasn't lunch, but nor was he the hungry diner. Essentially, because no one had been stupid enough to get into the water on purpose with this apex predator, it didn't feel threatened. There was really only one way to get to the bottom of this and I was about to become the guinea pig in the research.

We teamed up with Adam Britton, a world-class crocodile expert, and together we embarked on a journey that would take us back to the Okavango Delta of Botswana and then to the Northern Territory of Australia where we would also attempt our experiment with saltwater crocodiles.

The key to scuba diving with wild crocodiles is stealth and speed. Crocodiles are habitually shy, so you sneak up on sunbathing crocodiles on the shore, and as they slide into the water, you too must descend swiftly and reach the river bottom before the crocodile. It is a race against time to get through what is called the kill zone – the area of water from the surface to a few feet below – in which crocodiles prey on their target. If the croc sees anything splashing or floundering in the water, it will assume it is a vulnerable animal or an animal in distress and it will attack. But reach the bottom and you become invisible. Not literally; the crocodile will know you are there, but it will leave you alone.

In general we like to make sweeping generalisations in life. We love to stereotype, hype and hypothesise. What we love in particular is a common enemy, and predators, in all their forms, have an ability to enthral and horrify us. We consider ourselves the planet's top predators, and we are certainly one of the most dangerous, but occasionally nature bites back and reminds us of our vulnerabilities.

Nile crocodiles are responsible for more deaths than almost all other wild animals in Botswana each year. In a land where the rivers and lakes are used to wash, collect water, and water livestock, the number of people who lose their lives to opportunistic crocodile killings is unnecessarily high. The motivation behind our expedition was to try to understand more about crocodiles' underwater behaviour. If we knew more about where they nest or build their lair and predate, then we could help to reduce the number of deaths.

But first we'd have to get into the water with them, which was, quite frankly, terrifying. We had only a wet suit for protection – no cage or chain mail. Agility and stealth are key, and any protection would make you more, not less, vulnerable.

I remember one particular dive in the Okavango as if it was yesterday. I was sitting on the side of the speedboat in full scuba gear as we searched for a crocodile. My heart was pounding from the adrenaline. We spotted one sunbathing on the river bank, and it quickly slid into the water as we approached. That was our call to action. I held my mask to my face, said a prayer and fell backwards into my potentially watery grave. The world turned a murky shade of grey as I scrambled to reach the river bed. The current threatened to sweep me away, but using a small metal stick, I hauled my body through the current and into the ever deepening darkness.

Brad reached the river bed before me, and using an underwater listening device he guided me close to our target. Left a bit, right a bit, forward, and, before I knew it, I was face to face with a ten-foot-long crocodile. Just a few inches from one of the planet's most feared predators.

We undertook a series of increasingly bizarre experiments, from waving a ping-pong ball in front of its face to monitor its underwater vision, to emptying a small non-toxic ink cartridge into the water to establish the speed of the current.

As if that wasn't bad enough, over the following week we did dive after dive. And that's when the dreaded complacency settled in. Reassured by the consistency of our results, we became lazy. We weren't as careful as we should have been and we began to make mistakes. One particularly bad mistake almost ended in tragedy.

Our cameraman misjudged his entry into the water on this occasion, was caught on the surface and nearly got attacked by an agitated crocodile. Fortunately, he was saved by his camera, which he used in self-defence, but the incident meant we had to make a serious judgement call on whether to continue with the project. The reality was that despite our best intentions to change the narrative of the crocodile as a scary apex predator, we had inadvertently consolidated that fear with our own arrogant behaviour. These were wild, unpredictable animals, and it can be difficult to predict how a wild animal is going to behave all of the time.

Rightly or wrongly we carried on. We became more careful, cautious and respectful, and a few weeks later we concluded the Botswana leg of the expedition before heading to Australia to attempt the same experiments with saltwater crocs. With the benefit of hindsight, we were lucky. Our risk paid off and we made a beautiful film about a misunderstood creature. We were not only able to help change their reputation but also to take away some real science about the crocodile that could be used to reduce the human–croc conflict. On a wider point, could we call the experiment a success? And if so, was it worth the risk? If I'm honest, as a father of two I think it was too much of a risk, but then again I was lucky.

But how do you quantify success? First you need to have a measure of achievement. It's back to that flatline of mediocrity, with no peaks or troughs. Without those extremes in your life, it is hard to define what success is. As Winston Churchill once said, 'Success is the art of stumbling from failure to failure without loss of enthusiasm.' Failure has a habit of suffocating success, but it is also a vital ingredient in achieving it.

* * *

To be successful in the wilderness requires patience, tolerance, resilience and knowledge. Knowledge once came naturally to us, in the form of an instinct for survival. There are plenty of people and tribes and cultures who still retain those skills, but most of us have long lost them to the culture of convenience and haste.

By and large, modern urban culture requires haste. We race from one experience to the next, and it is all about being faster and quicker. Life has almost become a race. We are in a perpetual competition against the clock. We are either trying to beat it by cramming as many working hours into the day as we can, or we are trying to turn the clock's hands back so that we have more time.

And if we aren't trying to beat the clock, we are looking at ways to save time by other means. Fast food, fast culture. We are living in an era of Insta-gratification, we want things immediately without the effort or the time. Okay, we have been quite successful, but at what cost? Lethargy, obesity and intolerance. We are killing ourselves in pursuit of haste and speed. The wilderness, in all of her shapes and forms, can be brutal and volatile, but in her more passive state she can also slow us down to a natural rhythm. A rhythm like that of a slow, meandering river.

CHAPTER FOUR

ISLANDS

'I feel we are all islands in a common sea.'
Anne Morrow Lindbergh

'Wanted: volunteers to be marooned for a year on an uninhabited island.' We all have a defining moment in our lives. Something that changed everything. For better or for worse. Mine was when I answered an advert in a newspaper.

I was 23. I had just graduated from university in Latin American Studies but I was lost. I was living at home. No girlfriend. Mum and Dad were supplementing my low wages as I worked as a picture editor for *Tatler* magazine. Life was as urban as you could get. My only connection with nature was my daily walk through Hyde Park.

I was lost, but I was about to be found.

In life, is there destiny or do you make your own path? I think it's a little bit of both. Whatever the truth, the advert changed my life. Thousands applied, but eventually, against the odds, I was chosen as one of the 36 volunteers who would be marooned for a year on an island in the Outer Hebrides for a new genre of programme known as reality TV.

It was 1999 and the BBC wanted to celebrate the millennium with a ground-breaking social experiment. This was no game show. We weren't paid. There was no prize money, no voting off, no winner, no presenters on the island. It was simple, organic and natural.

I say 'against the odds' because I'm really not sure why they chose me. I was distinctly unremarkable. I'm not being hard on myself. It's true. I had no skills. I wasn't a farmer, a teacher, a doctor or a slaughterman, which was the reason a number of other volunteers had been cast. I wasn't particularly handsome. No six-pack or bulging biceps. No hilarious banter. No gleaming white teeth or botoxed face. I was just a shy posh boy without much going for me. Well, maybe that's a bit harsh but it's how I felt about myself. Incredibly, the producers felt differently, I would later learn that I was chosen as the token 'toff' to rile everyone else. Maybe my privileged education had paid off after all.

The premise of the programme, called *Castaway*, was simple. We would have a year to create a fully self-sufficient community. We would rear and slaughter our own animals; build a wind turbine and hydroelectric dam for our limited energy needs; build a school house to provide education for the children; erect polytunnels to grow fruit and veg; and we would install water piping to create a freshwater supply. The overarching plan was that over the course of 12 months, we would successfully transform ourselves from a mismatch of urban strangers into a thriving happy community. It was the most extraordinary year, marked with highs and lows, but it also cemented the foundations of knowledge in my relationship with nature.

Until that year, my relationship with the environment around me had been largely one-way. I was a consumer. In cities we take more than we give, but living so close to nature was an illuminating experience. I was never bored on that island. Not once. There were always places to go, tasks at hand, jobs to do and people to chat to.

In many ways, the experiment was more about community than it was about being 'castaway'. I loved the community life and it helped mould my steadfast attitude towards companionship, friendship and relationships. We were an eclectic bunch, and even back then, in the infancy of reality television, the producers had selected a group of individuals as much for their potential to come into conflict as their ability to succeed.

In those days, reality TV had an unaffected honesty and a natural ebb and flow to it. In the two decades since, the genre has changed beyond all recognition. Indeed, reality TV has become unreality TV. Producers no longer trust things to happen naturally. They want to know the start, the middle and the end, before it's even begun.

None of us was there for fame and fortune. We were there to experience island life; the TV show was incidental, a means to an end. And I think another point of difference was that *Castaway* was so much more than just a TV show. It was a way of life. A showcase for farming, for sustainability, and for community living. Even over and above the sense of community, *Castaway* was a showcase for a place. For island life. For the weather. For the

Western Isles. *Castaway* was about so much more than just the people.

It was Bruce Tuckman, the American psychologist, who hypothesised that a community would transition through four phases: 'Form, Storm, Norm and Perform.' *Castaway* proved it was pretty accurate.

The first month was a blur of whisky drinking, drunkenness and hangovers. The following months were beset with arguments, tantrums and people fleeing the island. But by the summer, we were getting on with life almost effortlessly, and by the time the final volunteers left the island in January 2001 we were thriving. It felt like we were being ripped from our surroundings, from the idyll where we had made our home.

Island life was beautiful, natural and organic. We awoke with the sun. In the summer we would rise at 4 am, in the winter we would sleep until 9 am. We worked with nature, not against her. The island was our teammate, not our competition.

The Outer Hebrides island chain is often described as one of the wildest, windiest and wettest places in the British Isles, but that was what made our experience so magical. The weather was one of the stars of the show. But the real show-stopper was the island of Taransay herself, with beautiful rugged cliffs and talc-white sandy beaches that often looked more like they belonged to the Caribbean than the Western Isles of Scotland. Taransay could be cruel with her wild weather, which made crop farming uncertain, but we, the islanders, had a beautiful, quiet respect for her. The wilds of a Scottish island were pretty alien to all of us at first, but we never felt threatened. Through time we learned to work with our surroundings rather than trying to battle against them.

So many subsequent reality TV shows have cast the island or the jungle as an inhospitable environment – pitting man against wilderness – but the beauty of our experiment was the synergy we had with our habitat.

For me, the most insightful part of island living was learning about cause and effect. In the urban, material world, it is often difficult to relate how our behaviour can have far-reaching ramifications, but here on this tiny island, we lived or died by our mistakes. We had to take responsibility for ourselves and for those within the community. So often we defer to others when it comes to decision-making or to help us when something is broken, but on that island the buck stopped with us. There was no boss. No king. No prime minister or president. We had made a bold decision to have no hierarchy. We would be leaderless. If I'm completely honest, I think it was the first time in my life I was forced to take charge of a situation.

So what did I learn during that year on Taransay? Let's begin with tolerance, particularly pertinent in these increasingly illiberal times. Society seems to be closing in on itself. It starts seething with outrage at the most trivial of things while at the same time letting great injustices pass unchecked. I think we have lost sight of what is important in life. The world in 2020 is increasingly volatile and polarised. It can be difficult to comment or make an observation on anything without risk of politicisation. Intolerance at one end of the spectrum and wokeism at the other end have all but obliterated the middle ground of 'tolerance'. The result is a swirling, malevolent hatred that often manifests itself online.

Tolerance is about being open-minded. It doesn't always mean conceding or capitulating, but it does require understanding. Perhaps empathy is a better word. Tolerate is what I do with Brussels sprouts. I eat them even though I don't particularly like them (cue angry online hate from the pro-Brussels sprouts mob). I know they are good for me and at Christmas they are a part of the festival of cheer, so I eat them, but that doesn't mean I will lie and tell everyone I 'love' them. On the contrary, I will ensure people know I am eating them even though I don't particularly like the taste of them, BECAUSE LIFE ISN'T ALL EDAMAME BEANS.

Life isn't always about perfection and idealism. It is full of things we don't particularly like. It isn't all raspberries and chocolate (my personal favourites), but Brussels sprouts and peas too (my least favourite).

The problem is that so many people expect perfection. The world really doesn't work like that. We won't and don't all agree. We have differing opinions, beliefs and views. It's what makes us a marvellously eclectic species. On Taransay, we had devout Seventh-Day Adventists, anti-vaxxers, people who were openly homosexual and others who were strict Catholics. We rarely agreed. We argued, but we got on with it. Tolerance and understanding were key to our success. Divide and you are conquered.

Castaway was interesting because it took a number of things away from our daily life. Consumerism and materialism were gone. We each packed a single trunk with our worldly possessions. If a coat ripped then you repaired it. If a hole appeared in your Wellington boot then you fixed it. There were no seasonal fashion trends. We wore the same clothes for the year and quite frankly it was brilliant. I wore the same Guernsey sweater for 365 days. As for my pants … I'll leave that to your imagination.

There were no shops. There was no jealousy of possessions. Books were shared, jackets swapped, presents were made from things found on the island. Consumerism was largely banished, and with that there came a sense of relief and satisfaction.

My wife Marina often asks me, 'Do you need it or want it?' It's a clever use of words. How often do we buy something because we simply want it? The real question is, do we really need it? Take consumerism and materialism out of the picture and you release a great pressure and attain a great liberty.

We were also shielded from news, information and the internet. I hate to admit this, but twenty years ago I didn't yet even have my own email account. The internet was in its infancy and social media wasn't even a twinkle in Mark Zuckerberg's eye. Mobile phones too were in their early stages: they had just progressed from the mobile 'brick' to the classic Nokia, but they were still for the few, not the many. And as for mobile phone masts on Harris and Lewis in the Outer Hebrides, forget it. We really were information castaways: no newspapers, no television, no radio, no visitors; we were allowed letters every few weeks, but apart from that we had no external interference. Imagine a life of no Radio 4 in the mornings, no free copies of the *Metro* on the London underground, no billboards advertising the latest fashion, no social media, no Kardashians, no Donald Trump, no Brexit or Megxit …

On Taransay, we were blissfully oblivious to the outside world. Of course, we heard snippets of news from passing fishermen who smuggled bottles of whisky onto the island. We weren't complete troglodytes, but information was rare and the result was that we could focus on ourselves and the island.

They say 'an Englishman's house is his castle', but for me the insinuation has always been that the English, as a nation of islanders, separated from the rest of the world by water, enjoy an element of isolation. If you were to equate island life to a kind of dwelling, it would be a castle with a moat and a drawbridge, separated from the world beyond. Island life creates isolation and separation. Islanders often retain a proud independence from the rest of the world. It's almost a 'them and us' mentality. Isolation doesn't just affect an island's human population; the flora and fauna all adapt to their individual environment. And it's amazing how quiet life becomes when you can finally silence the background noise and focus on yourself. Life has become a lot noisier since 2000, but two decades ago we reaped the mental health benefits of shielding ourselves from unnecessary mind clutter.

That's how I think of it. It's piles of rubbish in or on the brain. Have you ever had the feeling that you can't concentrate and you have lost control when there are too many piles of letters and unwashed clothes and stuff? Clutter clouds the mind.

Deal with the clutter and everything suddenly becomes clearer. Can you imagine what we are doing to our own brains and minds with the clutter that accumulates? Normally we would be able to filter out all of the detritus and unnecessary information in our dreams, but of course we are all far too busy absorbing information to dream. Most of us aren't even getting enough sleep.

On that island we slept an average of 12 hours a day. There were no prescribed work hours. Everything was voluntary. The pressures of work, money, tiredness and consumerism were gone. We were unburdened. That's not to say it was easy or always happy. Far from it. We filled some of those voids with vacuous local gossip. Small trivial incidents were often blown into all-out island conflict. Our *Castaway* experience was far from perfect, but there was a naturalness to that lifestyle that I haven't been able to replicate in the twenty years since.

I learned a great deal about people during that year. The most important lesson was, as I've already said, to respect others' opinions, even when they are different from your own. As a society we seem to have become increasingly intolerant. It's why we never seem to get anywhere. People are losing the ability to be flexible and adaptable. But if there is one thing you learn very quickly in the wilderness, any wilderness, it is adaptability. Adapt or die. It is the reason why so many have perished in the pursuit of their wild dream. You need to be chameleon-like in your environment. Take your arrogant, apex predator sentiments and values into the forest and you will be eaten by a bear. That is how nature works. You have to bend, shape and shift. It doesn't mean you can't remain true to who you are, but you can try and put yourself in another's shoes. You have to try and remember that it's not all about 'you', or 'us' or 'them'.

Another significant lesson I learned during that year was about our relationship with animals. As the son of a vet, you won't be surprised to hear that I longed to be a vet myself, but an 'N' in my A level Biology wasn't really going to cut it. But I still love animals. All animals. And yet I eat meat. Or rather, I ate meat.

The human–animal relationship is complex. On the one hand we invite one animal, the dog, for example, into our lives, our homes and in many cases, including my own, into our beds. Our dogs, these four-legged hairy beasts, have become a man and woman's best friend. I cried and mourned the loss of my dog, Inca, more than I did when my late grandmother passed away. Go figure.

We love dogs. We spend billions on them each year. And at the same time we are very happy to eat the flesh of a cow in a hamburger, or of a pig in a bacon roll or a ham sandwich. How did we figure that one animal is any more valuable, emotional, clever or sentient than the next? Living on that island, with livestock of our own, forced us to confront our own relationship with the animals in our care. We had cattle, sheep, pigs, horses and chickens. Of course, I volunteered to help with the animal husbandry and that would include the slaughter. Even aged 24, I had long acknowledged my own hypocritical view of animals and my own inability to confront the reality of having to kill the animals we eat.

Life has become so sanitised, many of us don't even know what meat we are eating. It is an abstract product that supermarkets have done their most to disconnect from the mooing or oinking animal we see on farms. It is, after all, in the supermarkets' interests to encourage us to buy as much meat as possible. Profit. Profit. Profit.

On the island that relationship was visceral. We fed the pigs. We watered the pigs. We played with the pigs. They became our friends. Then we had to kill them. I'm not really into killing. (That's an understatement. I hate killing.) But *Castaway* gave me an opportunity to confront the reality of my carnivore diet.

During the selection process for the project, the BBC had sent us to spend a few days with Lofty Wiseman, the television presenter and man behind *The SAS Survival Handbook*. Lofty had, and still has, a bushcraft camp in Wales and we spent a weekend learning some of the skills we'd need on Taransay. One of those was killing a chicken by wringing its neck. It was the first time I had ever tried to kill anything in my whole life. Not surprisingly perhaps, I bottled it and Lofty had to finish the job. Failure again. But now on the island, I knew it was time to confront death.

The first animal we ever killed was a pig. Colin was our designated slaughterman. A butcher by trade, he helped us build a slaughterhouse. We coaxed the pig into the small room using a bowl of food as bait. I'm not sure I can ever forgive myself for deceiving an animal and tricking it to walk towards its death. Once it had crossed the threshold, we grabbed it and strapped a noose to its legs. It squealed and flailed helplessly as we hoisted it up. Colin used a captive bolt gun to the head and then slit its neck. That's the sanitised version. I'm surprised I didn't go vegetarian there and then. I cried for hours.

This was but one tiny window into the plight of hundreds of animals that had been killed to feed me over the years. I looked at my own dog, Inca the Labrador, who I had taken to the island as my luxury item, and I imagined her hoisted up on that pulley. I had nightmares for days. But I ate the pork. I felt it was disrespectful to the animal to abstain after I had helped dispatch him. I have always equated meat-eating with our arrogant relationship with the planet, in that it's more about asserting our authority than caring for and respecting the living creatures around us.

Isn't it strange how we humans quantify the relevance and sentience of animals according to what we eat. In Great Britain, dogs, cats, rabbits and horses are all pets, while cows, pigs and sheep are sustenance. In France, rabbits and horses can be pets as well as dinner; while in Korea, dogs and cats are dinner, and in India, cows are holy. The only consistency is in our arrogant attitude that we will decide who we eat and who we pet. The outrage we vent on those nations that still kill and eat whales could easily be turned on us and our appetite for cows and pigs, the eating of which is frowned upon by other nations.

Who is right and who is wrong? Even at the age of 46, I'm still not sure. What I do know is this: that we don't really 'need' to eat meat. We certainly eat too much of it and the impact animal husbandry is having on the planet is hugely damaging. Hand-to-mouth slaughter is very different to industrial slaughter for commercial markets. The quantities required are staggering. For example, 80 billion animals worldwide are slaughtered for meat each year, according to a 2018 survey by Our World In Data. Confronting the realities of the provenance of our meat had a profound effect on our relationship with what we ate on Taransay.

I helped with every single slaughter during that year. I never pulled the trigger, but that doesn't abstain me from culpability. But if I ate meat, I felt it was only right that I participated in the act of killing animals for our dinner table. It's too easy to hide from the necessity of slaughter, to let others do the bloody work and pretend to yourself it never happened.

We had a wild herd of deer on the island. To help keep the population in check, the farmer had told us that we were permitted to take half a dozen of the older male deer each season. If we could shoot one, we could supplement our diet with venison. Except we didn't have a gun. But then one day, about seven months into the project, a rogue deer broke into our vegetable patch which we thought we had deer-proofed.

We didn't have time for a debate or a meeting, Colin grabbed his bolt gun (also called a humane killer), and a number of castaways, myself included, grabbed a fishing net and began to slowly corral the deer into a corner of the vegetable patch.

It was trapped. We moved in. Slowly. Silently. Breathlessly. I could hear my heart pounding. Fear. Excitement. It was a primal instinct. It horrified me but it was there. We jumped and pinned it to the ground with the net.

BANG.

It was dead. We were stunned. We couldn't quite believe we had caught and killed a deer. The community was divided. The vegans were disgusted. I was horrified.

We ate every last sinew of meat.

It sparked a lively conversation. Which is fairer? To kill an animal like the deer, that has lived a wild, free existence, or to rear pigs in captivity for slaughter? If you are a vegan or a vegetarian then you'll probably argue that they are both awful and you shouldn't kill either. But if you include meat in your diet, then it is a complex question.

In subsequent years, I have come across rural folk who hunt for their plate. I have spent time with bear-hunters in Alaska, moose-hunters in Canada, elk-hunters in Sweden and eland-hunters in South Africa. Each time I am horrified by the hunt. The guns. The blood and the butchery. But, I ask myself, which is worse? Hunting for the free-range animal that has known freedom all its life, or rearing the factory-farmed animal that has never known life in the wild? Twenty years later and I am still equally conflicted. It is one of the reasons I am now largely vegetarian. When I say, 'largely', I don't like the notion that we must define ourselves. There are times and places where it is socially wrong, rude or difficult to say no. Hypocritical, maybe, but then isn't life full of compromises?

Living and working with animals was a big part of the *Castaway* experiment, but, like I've said, it was our relationship with the island itself that was life-changing.

Without any outside distractions we got to know our landscape well. Over the course of that year I must have walked hundreds of miles along Taransay's beaches and her craggy cliffs. I would watch the weather as it rolled in off the Atlantic Ocean. I observed the roiling, frothing water as storms raged all around us. We became close to the wildlife. The seals and the birdlife. The otters and the basking sharks.

I'm proud to say that we had a pretty insignificant footprint on that island. We lived 'softly'. It was the year I became a conservationist. Acutely aware of cause and effect, I understood that whatever we did would have an immediate impact on our lives and on the health of the island.

The year was significant in many ways. It really did change my life. It would lead to untold opportunity but it also helped rebuild that shattered confidence from my early childhood. The new confidence did not arise from success or fame but through mental well-being. We were living simple, natural, organic lives free from complication or outside interference. My self-esteem flourished and I finally felt secure in my own skin, and comfortable with who I was.

Leaving that island was pretty traumatic. The psychologist working with us, Cynthia McVey from Glasgow Caledonian University, described it as a bereavement. We loved that island but our time was up. For me, leaving felt a little like regressing, a step back, as we were helicoptered from splendid isolation and solitude back to harsh reality.

In my case, that was back to London, back to Mum and Dad's home, back into my childhood bedroom. It was like sliding down the snake in a game of snakes and ladders and I was back at square one.

I may have lost some of my inner confidence but I had found fame. Or it had found me. The thing about reality TV fame is that it is very shallow. There is no substance. I was famous for being the posh boy on *Castaway*. I had been given my fifteen minutes and the stopwatch had started. If I wanted any longevity in my TV life then I would need to put down roots and give myself a solid platform from which I could build something of substance. When life throws lemons at you … make lemonade.

A lot has happened since *Castaway*, and there have been a number of lessons that people have shared with me over more than two decades, but at the heart of everything I've gone on to do is that life lesson on Taransay. I wish everyone could experience such a meaningful, life-changing experience themselves.

The experience cemented my own relationship with the Outer Hebrides. I have travelled to hundreds if not thousands of places since, and none of them has had such an acute impact on me. I have returned both to the Western Isles and to Taransay many times over the years. I even took Marina and our two dogs there for our honeymoon where we slept in a single bunk bed in the old school house.

At the beginning of 2020, I returned to the island on a pilgrimage to mark the 20-year anniversary of *Castaway*. I was struck by the timelessness of the Outer Hebrides. Everything on the neighbouring islands was the same. It is almost like being stuck in a time warp. I don't mean that in an unsophisticated kind of way, because what the island offers is a healthier way of life. The speed of city life has become frenetic. It can be difficult to keep up. People are finding it hard to switch off, and yet the answer for many is almost on their doorstep. Just a short ferry journey from mainland Britain is a series of islands that offer respite and recovery. Imagine if all those burned-out souls, consumed by social media and caught in the loop of consumerism, could step off that wheel into the real, working landscape of the Outer Hebrides.

When the newspapers first reported breathlessly on 'the bravery of the pioneering social experiment', many people on the surrounding islands pointed out that 'the bravery of the castaways being marooned for a whole year on a Hebridean Island' had been practised by tens of thousands of people for many centuries. They were quite right. There was nothing brave or unique or pioneering about our project, but it took 36 individuals out of their comfort zone for a year and there aren't many people prepared to do that with their life. It comes back to comfort and conformity. We are creatures of habit and sometimes it's difficult to break that status quo.

The year certainly took me out of my comfort zone. It taught me diplomacy and liberalism. It taught me to be open-minded and tolerant. It taught me to confront the knock-on effect of my lifestyle on living things and on the environment, and it taught me the importance of community. You see, as much as that cast-away year was about the island, it was also about the people. The community made the project what it was. Without the people, the experiment would have failed. It took teamwork and collaboration to succeed. We could never have achieved the self-sufficiency without the numbers.

I believe a great deal in the notion of community. I see it in working action all over the world. The small body of neighbours and extended family who look out for one another. We in the West have largely lost sight of community and neighbourly values. Of course there are some beautiful examples across the country of strong communities pulling together, but certainly in the urban theatre it has largely been lost. Society has become a little more selfish. It's survival of the fittest … or the fastest.

The coronavirus crisis of 2020 has brought home the stark reality of our reliance on others and it has also acted as a reality check on our previously fragmented society. The coronavirus isolation has given us all time for reflection. It has been a period of forced self-reliance that has made us more resourceful, less wasteful and, largely, a little kinder. It has created a sense of community born through isolation. Coronavirus lockdown has given us all a tiny sense of what isolation is really like. I always thought that in the event of a global crisis I would return to the wilds of the Western

Islands, and then, when it happened, I found myself isolated with my family in Oxfordshire, far away from the wild isolation of the Western isles.

Community fosters love, kindness, positivity and spirit; without it, you are in danger of disappearing down a wormhole of hatred and spite. Community lifts where solitary anonymity sucks you under.

I have a theory that I have held for a number of years on the subject of fame, community and social media. If you agree with my sentiment that people need community in their lives, then perhaps the addiction to social media and the clamour for fame are the result. You see, with fame and social media you create a wider community. A sense of place. People comment on what you are wearing, what you are eating and who you are dating. Just as small communities revel in the tittle-tattle of curtain-twitching gossip, social media and fame create an artificial wider community.

This social media 'community' has become quite toxic, especially when it comes to politics. Suddenly those who have felt ostracised or abandoned by society have a home, a place of kindred spirits that often act as an echo chamber to dangerous thoughts. The result is ever more ridiculous conspiracy theories. Fake news and angry Twitter and Facebook mobs: it isn't really healthy for anyone. Real community is different. It is how some of the most successful tribes live in some of the remotest corner of the world.

A collaborative and sharing culture is less harmful to the planet, and perhaps it's one of the reasons we were so successful as a community on Taransay. There is no way a smaller community could ever have achieved so much in so little time. Almost everything I have done in my life subsequently has the DNA of that unique year-long experience. I like to think of it as my wilderness apprenticeship.

Island living is not at all uncommon; almost all ocean-fronted nations have their own isles. You may be surprised to hear that there are 6,289 islands in Great Britain. Sweden has 221,800 islands. I don't think anyone has ever been able to count the number of islands across the world, but it must be in the millions. Many of them are uninhabited, but plenty of others are home to communities of varying sizes. My year living on Taransay may have solidified my love of islands, but I have been captivated by the notion of island isolation ever since I was a boy.

I was seven years old when I visited the Isle of Eigg in the Inner Hebrides, an island rich in history. I visited when it was owned by the Laird Schellenberg, who was eventually driven off the island by the tiny community who in 1997 started a crowdfunding campaign to raise the £2 million to buy the island for themselves. It became a beacon of hope for other communities hoping to claw back control from island lairds.

My sister and I spent a week on the island with friends. There was something so exciting about the intoxicating mix of isolation, remoteness and intimate community. Everyone knew everyone. We would walk from house to house and visit islanders without prior planning. We'd just turn up. It was so different to life in London. We picked vegetables from the veg patch and collected eggs and meat from one of the farmers and fresh fish from fishermen in the little harbour. I suppose if I think back, it was my first experience of self-sufficiency and I loved it.

Many find the remoteness of islands intolerable, while others are seduced by the romance of the isolation. For me, islands have a unique intensity, a captivating ruggedness. You can't just drive to the shops; electricity and heating can't be guaranteed; you are at the mercy of the weather, but what you may suffer in inconveniences you make up for in privacy and seclusion.

Shortly after I left *Castaway* 19 years ago, I set out on my own voyage around the UK coastline in search of an island idyll. I wanted a remote, wild place of my own. I had been seduced by the notion of 'islandism'. I had become addicted to the happiness that islands can bring with their simplicity. I put in an offer for a tiny island in the Summer Isles of Scotland near Ullapool, but I was gazumped by a wealthy landowner and decided to see if there was another way of finding island life.

I visited Bardsey Island in Wales and Muck in Scotland, both of which were looking for caretakers. Bardsey recently advertised for a family to move to the island, offering a £24,000 annual salary. Shortly after *Castaway*, I visited the island myself and managed to help secure the job for some former castaways who had become equally addicted to island life.

One of the most fascinating groups of islands is the St Kilda archipelago off the Western Isles of Scotland. Situated 40 miles west of the Uists and separated from the rest of the world by a stormy ocean, isolation dominated St Kilda life. As late as the nineteenth century, islanders could only communicate with the outside world by lighting a bonfire. This was eventually replaced by the famous 'St Kilda Mailboat', which consisted of a piece of wood and a small bottle attached to the inflated bladder of a sheep. A letter was placed inside the bottle, and when the wind was in the right direction, the 'mailboat' would be released. Two-thirds of the letters were eventually found in Scotland, or less conveniently, in Norway.

The isolation spared them the 'evils' of life elsewhere. One early visitor to the islands in 1697 noted that the islanders were 'happier than the generality of mankind as being almost the only people in the world to feel the sweetness of true liberty'. Can you believe such an observation coming from the 1600s?

In many ways, the sentiments of that early insight provide the narrative of this whole book. How many of us really know the 'sweetness' of freedom? The pure freedom that comes from escapism and isolation? Remoteness and isolation foster a unique, unaffected happiness.

Despite hundreds of years of occupation, the islands were finally evacuated in 1930 after the death of four men from influenza, and the death of a young woman, Mary Gillies, who was taken to the mainland for treatment for appendicitis from which she later died. Islanders became concerned about the vulnerability of their isolation and their ability to receive prompt medical assistance, so the decision was made to evacuate. The houses, the farms, the church and the school house were all abandoned as the islanders fled. The livestock was rounded up and sold, while the working dogs were sadly drowned and the islands were forsaken.

St Kilda is as difficult to reach today as it was to leave a hundred years ago. Unless you have access to a helicopter, the only way of getting to the archipelago is by open boat journey and hours navigating the often angry waters of the North Atlantic.

I have been to the islands twice, and the first time nearly broke my ocean spirit. I was actually on my way to an even more remote place, Rockall, a further 160 miles west of St Kilda, aboard a Danish sail boat called the *Eda Frandsen*. With her heavy mast and sail, she had listed alarmingly in the violent storm that engulfed us as we headed into the North Atlantic. I have rarely been so seasick and uncomfortable at sea.

The gathering storm forced us to look for safe anchorage and we eventually pulled into the comparative safety of the bay in front of the abandoned community on Hirta, the main island of St Kilda. Now used by QinetiQ, the military contractor for unspecified reasons, the island was home, for a period, to Britain's most remote pub, the Puff Inn.

Walking down the main street of that abandoned community, into the buildings that had once been home to hard-working, happy families, into the school house where I could imagine the children reading at their desks, was an experience I will never forget. It had slightly more sophisticated echoes of my own year isolated on Taransay, but this had been real life. This had real history and heritage.

I would return several years later with Norman Gillies, the son of Mary whose death had precipitated the evacuation. He had tears in his eyes as we wandered through the deserted settlement. Places that have been abandoned prematurely have a profound sense of sadness.

Islands become a microcosm of a wider society. As an island nation ourselves, we are islanders, but even beyond the shores of the mainland, we are a nation of many. Not many people realise it, but there are more than 6,000 islands in Britain, and over the years I have visited hundreds of them in my pursuit to understand the subtext of what it means to be an islander.

We seem to be fascinated by the stories of those characters, real or imagined, who, for different reasons, have found themselves 'castaway' or abandoned on islands. There was of course the flawed hero of Daniel Defoe's *Robinson Crusoe*, and then there were all those troublesome boys in William Golding's *Lord of the Flies*. Tom Hanks experienced his own kind of hell in the 2000 film, *Cast Away*, while Lucy Irving found hell of a different kind on a desert island in her 1983 autobiography, *Castaway*. The island is an effective setting for analysing the effect of isolation on a person. Islands also force people to be more resourceful, making due with what is available, and more flexible, dependent on their geography and weather.

Island cultures tend to be rich in folklore tied to their nature. Where urban society controls the wild, island life is determined by the rhythms of nature. The natural boundaries of islands have helped shaped their histories.

But what about those island communities that, for various reasons, have been forced to abandon their homes? Although the decision to leave the remote Scottish island of St Kilda was apparently unanimous, it must have been an incredibly difficult one. And in other parts of the world, communities have had no say in the matter and have been forced to leave, against their will.

* * *

In 2001, I set off to write my very first book. It was a pretty epic journey on all counts: the 100,000 words for the book itself, which was a pretty formidable challenge for the dyslexic boy who failed his English exams in school, but also the 100,000 miles that would take me around the world several times. From Pitcairn to St Helena, I travelled by boat to some of the most isolated communities in the world, but it was one, Tristan da Cunha, a British overseas territory, that was particularly fascinating.

Officially classified as the most remote inhabited island in the world, Tristan, with its 300 inhabitants, lies around 1,500 miles off the west coast of South Africa in the middle of the South Atlantic, between Buenos Aires and Cape Town. There is no airport, no deep-water harbour and no regular boat service.

The island itself soars up from the ocean as a volcanic cone 2,000 metres above sea level, and it occasionally erupts. It was one such eruption that led to one of the most remarkable and unhappy evacuations in recent history. That happened in 1961, when the entire population of 268 inhabitants was evacuated 6,000 miles away to the county of Surrey in southern England. For almost everyone, it was their first taste of life off the island. Initially forced to live in wooden huts at a disused army camp, life in England proved harder than island life for the Tristanians. The islanders lacked immunity to common infections and illness became rife. They became bitterly unhappy with life in the UK. They lamented the loss of the isolation that had made them happy. They were eventually moved to permanent homes in nearby Hampshire, but after several desolate years away from their home, the Tristanians eventually voted 148 to 5 to return to their island.

Another overseas island territory of the United Kingdom that was evacuated was the Chagos archipelago – situated in the Indian Ocean between Tanzania and Indonesia. The islands had been commandeered by the British government in the 1960s and handed over illegally to the Americans. The islanders were deported.

I will never forget wondering through the abandoned settlement. The tropical vegetation had suffocated many of the buildings, strangling and reclaiming the tiny village for itself. There was a deep sense of loss as I crept through the overgrown graveyard and into the ghostly houses. The place was not whole without its rightful community.

The story of the Chagos islanders' treatment at the hands of the UK government is one that should shame us all. It began with the disintegration of the last pink bits of the Empire. While Britain ceded independence around the world, she held onto strategic islands in the Pacific, Atlantic and Indian Oceans. When independence was granted to Mauritius in 1968, it was agreed that Britain would hold onto a tiny group of islands known as the Chagos, and they become known as the British Indian Ocean Territory.

Around the same time, the United States, worried about Russian expansionism, wanted a base somewhere in the Indian Ocean. Keen to foster relations with America, the British government held a secret meeting in London at which the islands were declared 'closed', and in a series of letters, shown neither to Parliament nor to the US Congress, the islands were 'leased' to the US for 50 years with the option of an extra 20-year extension. The deal was struck on the understanding that there would be no 'population problem', with Britain agreeing that the entire island chain would be 'fully sanitised' and 'cleansed' of human life. In exchange, Britain would get an $11 million subsidy on the submarine-based Polaris nuclear deterrent system from the US.

The 1,800 Chagossians were evicted and sent into exile. Their British citizenship was revoked and they were forced from their homes. Their 600 dogs were gassed to death and they were sent to live in squats and even prisons in nearby Mauritius and the Seychelles. In return, each islander was offered £325 for the loss of their islands, homes, livelihoods, and history. Overnight they had been made non-people. Britain had tried to erase them from history. Their islands became home to one of the most notorious American military bases in the world, Diego Garcia, dubbed the 'Guantanamo of the East'.

Back in 2001, I visited the island group. Technically it was an illegal visit. I sailed into the archipelago from the Maldives. I was able to walk through the derelict, abandoned villages that had been forcibly evacuated in such haste. The houses had been left exactly as they were, like a graveyard overgrown and untended. What shocked me more was the temporary community of sea gypsies living within the ruins. Where the islanders had been denied the chance to return, Westerners were squatting in their homes.

I have fought for the islanders' rights to return ever since. In 2000, the islanders won a high-court ruling and the then foreign secretary, Robin Cook, agreed to return their passports and their citizenship. Many of the islanders came to Britain and settled in Crawley to await their legal right to return. But on the death of Cook, Jack Straw reneged on the promise and refused to let them return, lest it anger the Americans.

The islanders have spent more than 50 years fighting for the right to return. For nearly 20 of those years I have been fighting their corner, meeting ministers and arguing their case. It is a story that makes me ashamed to be British.

In 2010, the islands were declared a marine protectorate. The conservation world was surprised at the haste and ease of the declaration. It has since transpired that the protectorate had nothing to do with protecting the environment, and everything to do with protecting the military base and ensuring the islanders would never return.

'This has nothing to do with the environment,' said the Mauritian High Commissioner to London. 'They want to prevent islanders from going back and to keep these islands forever. But we are not going to let this go.'

The Chagos islanders have been fighting for their right to return ever since – a return to the islands they still call home.

The remote wilderness cushions us from the superficial façade of modern suburban life in which we hide behind an illusion. As we are increasingly drawn into a virtual world, we are tempted to construct fantasy lives online. We project a notion of happiness that is often built on an edited lie. We are living our lives in a hyper-real world where it can be difficult to distinguish the line between fact and fiction. The line has become blurred as we search for the happiness that continues to elude us. With this hyper-reality dominated by instant gratification, in which we can buy a new sweater or even a new car from our beds in the middle of the night with a swipe of our fingers, we have lost all sense of provenance.

Oblivious to our own environment, it is perhaps not surprising that we have come to neglect it. But what remoteness, isolation and loneliness – all aspects of island life – soon teach us is that happiness is always possible there, you just need the tools to recognise it. Isolation gives us the patience and time, and integrity. We are forced to value our resources. We are forced to be more resourceful with our finite supplies. We are aware of cause and effect and the result is that we ultimately become kinder. There is nowhere to hide on an island; stripped to our basics, we become a better version of ourselves.

I still remain obsessed with islands. Whenever I spend time on islands I get a sense of calm fulfilment that I rarely get back in society. They say that no man is an island. The phrase expresses the notion that human beings do badly when isolated from others and that we need to be part of a wider community to thrive. But most islands thrive because of their own sense of community. We can have isolation and remoteness and still flourish.

A calm, modest life brings contentment and happiness. Away from the pursuit of success and the jealousy of comparison, we are free to be who we are, not what society wants to mould us into. There is a great beauty to island isolation (where you are there in the moment and not constantly striving for something more), not to be confused with coronavirus isolation (where you are stuck in the moment and desperately wishing yourself away).

No man is an island, but no island is the same without man. Islands can be restorative. They force us to be more caring and tolerant. They force us to be more self-reliant, and most importantly, they can be the source of our happiness.

CHAPTER FIVE

THE JUNGLE

'And into the forest I go to lose my mind
and find my soul.'
John Muir

Life can be hard but it is hardship that makes you and defines you. Some of the most inspiring and influential people on earth have overcome hardship to achieve and excel. There is no success without hardship. It is the divine, sublime, reality of life. You need the colours, contours and textures to appreciate what is ahead of you.

But you never really appreciate what you have until you take yourself out of your comfort zone and endure hardships. And of course hardships are relative. A miner, working in the heat, deep below ground, will endure hardships that are very different to those who endure the hardships of living in a remote cabin in the woods; in the same way someone on the bread line who requires help from a foodbank in Manchester will know and endure very different hardships to someone who lives in a refugee camp in Sudan. None of them is necessarily easier or harder. It is all down to comparative circumstances. Some of the toughest, most hard-working folk I know are those who have suffered and lived through hardship. The experience thickens the skin and strengthens the resolve.

There are few places as testing as the jungle, which can be one of the most difficult, unrelenting environments to live in. It is a place that exposes our weakness, and tests our fragility. It is a place that bites and stings and poisons and scratches. It is an environment that exposes our lack of knowledge or planning. Go into the jungle without basic skills, local knowledge and training and you won't be coming home.

I remember reading the story of Colonel Percy Fawcett, the British surveyor and explorer who went into the jungle and never came out. In 1925, Fawcett set off to Brazil to find the legendary Lost City of Z. He left strict instructions that if he failed to return, no rescue party should be sent to recover him.

Just a few months later, he and the rest of his expedition disappeared, forever. There is ongoing speculation as to whether they got lost and died of dehydration and starvation or whether they were killed by one of the local tribes. Whatever the truth, the story of Fawcett has only enhanced the mystery and fearsome reputation of the jungle.

The first time I ever visited the jungle was when I was 18. I had just failed my A levels. I worked in an ice-cream parlour for five months and earned enough money from selling ice-cream sundaes and banana splits to buy myself a one-way ticket to Brazil. Inspired by watching Tarzan on television as he swung from tree to tree, escaped quicksand and played with leopards, and having seen and read countless depictions of the jungle in films and books, I wanted this great place for myself.

The Amazon rainforest is the largest jungle on earth and it seemed the most likely place to experience the real 'forest'. I flew into Rio de Janeiro. Got mugged. Got on another plane and flew to Belém, at the mouth of the great Amazon river. I will never forget the first time I went to the harbour to look out across the water. What I found was not so much a river, but an ocean. At its widest, the Amazon is a staggering 120 miles wide. For a boy who had grown up in Central London with the River Thames as a reference point, it was breathtaking and overwhelming. The port was bustling. Filled with hundreds of boats and ships of all sizes. I wandered from vessel to vessel until I found one that was heading to Peru, nearly 3,000 miles upriver. I negotiated a ticket, slung my hammock and headed into the jungle to become Tarzan.

For four long weeks we slowly chugged our way up the river. The rainforest was a mere green blur on the horizon as we meandered from port to port to drop off the cargo of biscuits to various communities along the way. Not knowing any Portuguese, I found it a pretty lonely journey, despite the hundreds of other commuters that were on board. I would sit on the upper deck as we pulled into each port. Trucks and cars and thousands of people milled and bustled and hustled. And then we would set off again, the jungle remaining tantalisingly in the far distance.

By the time I reached Benjamin Constant on the border with Peru, I had hardly seen a tree, let alone the 'jungle'. The most wildlife I had seen had been the mosquitoes that feasted on my blood each day, the bats that terrified me by hanging among the hammocks and the piranha that featured in our dinner each night. I had spent a month here and yet the jungle had eluded me and remained a distant mystery.

That was until I found myself in the middle of a primary rainforest for the first time when I went on an expedition to the jungles of Guyana on a trip to Kaieteur Falls. This was the first series of *Extreme Dreams* – a BBC programme in which I took a team of everyday folk on life-changing journeys around the world. Each of my teammates had experienced a shattering event: illness, bereavement, loss, a mental health crisis. They had each been damaged in one way or another and the TV series was my way of helping them with their recovery through expedition and the wilderness.

It was 2006, I had just rowed the Atlantic and was still flush with the feeling of invincibility. We spent a few days in 'jungle school' in which we learned the lessons necessary to endure the conditions, and then we were released into the rainforest. It really was a baptism by fire. It was both awful and brilliant.

My memories of that expedition are deeply ingrained. I can picture myself wading up to my chest through the brown, muddy waters. My rucksack, wrapped in a waterproof bag, is trailing behind me on a tether, floating like an aquatic dog. Stumbling over submerged tree roots and weaving around the flooded limbs of trees, the sun dipping onto the horizon and the last smudges of pink and red obscured by a blackness that is spreading its tentacles like ink on blotting paper. I am in the middle of primary Amazon rainforest, the Guyanese jungle of South America. It is my first experience of the jungle and I am well and truly out of my comfort zone.

The jungle can be one of the most testing landscapes on earth: an oppressive mix of dense trees, spiky flora and venomous fauna, tropical heat and torrential monsoon-like rain, and hard-to-navigate topography with its rivers and swamps. There is no escape from the jungle. It is all-consuming. It clings to you and gets into the very pores of your skin.

We are beaten by the daylight and are wading through the flooded jungle in darkness. Half the battle of the wilderness is in the mind and in my head: I imagine the crocodiles, piranhas, snakes and leeches beneath the surface of the water. In the evening we find a dry patch of forest to string up our hammocks and make a little fire to cook some food.

Everything is a battle. You need to clear the sharp, spiky vegetation from the floor and then make sure there are no ant nests on the trees. You only need to make that mistake once to ensure you never do it again, especially when you wake up with hundreds of stinging ants all over your skin. You need to check above your head for any dead branches from the trees that might fall in heavy rain, and then you string up your hammock, mosquito net and 'basha', or waterproof cover, which protects you from those frequent downpours. Then you face the difficult task of starting a fire using the sodden wood. And once you have finally got the fire going, you want to keep as far from it as possible to avoid sweating more than you already are.

You are perpetually drenched in the jungle. Clothes, boots and socks are all soaking, like you have just jumped into a pool, or in my case, waded chest deep through a flooded river. The key to survival is having one set of clothes that are comparatively dry and which you keep carefully wrapped in a dry bag, within your rucksack. Peeling off your wet clothes, it's vital you try to dry your boots next to the fire, making sure you use sticks to upend your boots to ensure no scorpions or spiders climb inside during the night. Still sweating and wet, you pull on your comparatively dry clothes before sliding into the hammock for a restless night's sleep.

Then there is the noise at night. I'll get into trouble for calling it that. Every soundman I have ever worked with has castigated me for getting into a muddle over 'noise' and 'sound'. There is a distinction. Noise is unwanted, sound is beautiful. After all, they are not called 'noise-men'. Sound is something we can enjoy hearing, and believe me, the sound of the jungle is unlike anything you will ever hear anywhere else on earth. The wildlife is like nature's orchestra, creating a melody of sounds so rich and bold and vibrant and loud that it can be difficult to sleep.

I lie in my hammock and listen to this crescendo as frogs and toads and crickets and birds and hundreds of unidentifiable creatures, both great and small, join in with nature's symphony. And then, as I lie there, struggling to add an animal or insect to a particular sound, I begin to hear noises out of context: house alarms, pneumatic drills, garden strimmers, leaf-blowers, car horns and the whoosh of trucks.

Next I hear the gentle call and cry of babies and young children: their sniffles and hiccups, coughs and sneezes. It is as perplexing and confusing as it is magical. Eventually, I fall asleep, lulled into my dreams by the sound of the jungle orchestra.

On our first night in the jungle we awoke to the sound of

cracking bones. I lay there frozen in my hammock, too scared to move. In the morning, we saw in the middle of the camp the remains of a wild boar, and dozens of footprints all around our hammocks. A pair of jaguars had finished off their dinner inside our camp, and all we had was a thin silk hammock for protection.

Each morning, I peeled the leeches from my skin and flicked the giant spiders from my mosquito net, before changing into my soaking-wet jungle clothes. There can be little that's worse than pulling on wet socks and pants in the morning.

After two weeks, we emerged, shattered, scratched, skinnier and covered in bites, and I swore I would never return to the jungle. Little did I realise that I had inadvertently taken a part of the jungle with me …

I got back to London with a stinking rucksack filled with damp contents. Worried about the smelly mess within, I left it in my garden for a few days until I had the strength to unpack and wash everything. As I opened the rucksack, a huge tarantula, the size of a human hand, crawled out and disappeared into the undergrowth. I called the keepers at Longleat Safari Park, who reassured me it was unlikely to survive a British spring. To this day, I imagine a Guyanese tarantula roaming the streets of Notting Hill …

It wasn't my last visit to the jungle. My loud, vocal dislike of the rainforest environment was too much for TV execs to resist; they wanted to see me out of my comfort zone and just a year later I found myself in another jungle, this time in Papua New Guinea. Once again, I was filming for *Extreme Dreams* and this time we were trekking along an old World War II battleground called the Black Cat Trail, a parallel trek to the infamous Kokoda Trail, which was the location of one of the most critical and bloodiest battles of the Pacific war in 1942. Both trails were used by the Japanese to try to capture the capital, Port Moresby, from Allied forces via an overland route after they had failed to do so from the sea.

Within moments of being in the rainforest all my memories came flooding back: the leeches, mosquitoes, insects and poisonous plants. Hot and humid, with the risk of endemic tropical diseases like malaria, it is a hostile, uncomfortable place. What marked this jungle landscape from that of the South American one was the impact of the war along the trail. The shattered remains of abandoned aeroplanes, ammunition dumps, large guns and tanks littered the route. At night we found ourselves stringing our hammocks to the trees of old camp grounds once occupied by Allied forces, the ground still littered with the odd remnants of dog tags, army uniforms and spent bullet casings. It added a heavy sadness to the journey. The jungle had been as gruelling and soul-destroying as the enemy itself. The Black Cat Trail was also harder than Guyana on a number of other levels. It was more physically demanding, being more up and down, so the trek took more out of your body; but it was also much tougher because of the leeches.

I don't mind admitting that I really dislike leeches. I had first come across them as a youngster in Canada. My father had taken my best friend and me on a camping holiday in Algonquin Provincial Park in Ontario. I was 13 and I found my first blood-sucking leech on my foot. I freaked out, overcome by a feeling of revulsion. But that encounter was nothing compared to what we faced in the rainforest.

It was impossible to keep the leeches at bay. The same goes for the jungle itself. If you try to defeat it and keep it away, you are simply fighting an exhausting, losing battle. Those that struggle the most in the jungle are those who try to cling onto control. So each day I would remove tiny black blobs from my legs and arms. They would crawl into my hammock at night and silently begin sucking on my blood. But there is no control within the jungle. You have to abandon power to the trees, and once you have let yourself free, the jungle starts to become familiar and you can start to enjoy the landscape more. It's a beautiful thing to be at one with nature, particularly when it is, on the face of it, hostile. So while I never felt completely at home in the jungle, I began to understand it. And the lesson is to work with it, rather than fight against it.

For the third series of *Extreme Dreams* I headed back into the jungle. But the conditions were far from good. We were in Latin America, in Ecuador, a country I love dearly, at the foot of Chimborazo, at 6,300 metres the highest mountain in Ecuador and a formidable peak to climb in the Andes.

For me though, it was the beginning of much more than just climbing Chimborazo. I would be in the Americas for 12 weeks on four consecutive expeditions that would continue into the Canyonlands of Peru before descending into the Amazon rainforest and finally rising to the arid desert plains of the Atacama desert in Chile.

While I had previously spent years away from home, now I had a new wife and a new home, and suddenly 12 weeks of hardship felt like a battle. The wilderness for some reason had lost her allure and I had probably lost the battle before it had even begun. I can remember lying in my small leaking tent at the foothills of the Andes. It was day one of many and I was already feeling overwhelmed by the scale of the challenge ahead of me. It had been several years since I had rowed the Atlantic and first experienced the sensation of helplessness. Helplessness can be overwhelming. Surrender to it and it spirals into anxiety and then, out of control, you will disappear into your own vortex of despair.

In that tent, my heart was racing with panic. I just didn't want to be there. Although I was in the midst of my dream job, paid to climb mountains and explore jungle, I felt claustrophobic and burdened by the feeling of oppressiveness. This was my expedition. My journey. My show. I was leading much less experienced people on a life-changing adventure and yet I couldn't cope myself.

I have often wondered why I felt such despair when I should have felt elation. I think it was because I lost control of my mind. The loud inner voice had somehow hijacked my thoughts and I was being poisoned by hopelessness.

So how did I come back from that? It would have been easy to give in to that helplessness. I could have faked an illness or an injury (I genuinely considered this at the time), or I could have just walked away from it.

It is so easy to fixate on the negatives, to burden your emotions with the worst-case scenario and become that glass-half-empty person. It was too easy to focus on the never-ending rain that lashed at my tent, the cold water running down my neck and the thick mud that would clog my boots. I could still feel the slippery trail beneath my weary legs during the short, sleepless nights in a waterlogged, muddy sleeping bag. I obsessed about my rucksack, which would hold all my worldly possessions for the next six weeks.

I endured the first night, but the feeling of doom and gloom wouldn't shift. For the first time in my adult life, I was overwhelmed by negativity. It would be wrong to call it depression, but it was a metaphorical dark raincloud that wouldn't shift, even once the real rainclouds moved on.

We failed to climb Chimborazo. It wasn't entirely my fault. I think the ambition of getting a team of novices to such great heights with so little experience in such a short time was generally ill-considered, for which I must partly accept the blame, but I also think my mindset at the time didn't help matters.

It was a bitter pill to swallow. My teammates were gutted. They still are. The series was meant to be recovery through expedition, but our failure to reach the summit had led to a very public failure on national television that only amplified the disappointment. While the first group of teammates headed back to Britain, I felt despondent and culpable for their failure, but I had to carry straight on to another expedition, this time into the Canyonlands of Peru. By now I had accepted the fact that I was in this for the long term. But I still found it difficult to shift that metaphorical raincloud that hung over me.

Why did I feel such a continual feeling of helplessness? I think it was because I wanted so much to control the narrative. We were making a television show. I so wanted to prove my theory that the wilderness can help us all. I was so desperate to be right. Desperate for success. Desperate to be happy.

The expedition through the Canyonlands was largely successful and I managed to regain control of that feeling of helplessness. But then we entered the jungle again. And the helplessness returned. Sleep-deprived and exhausted from the two previous expeditions, I was now heading back into the jungle, that most challenging of environments, with no sense of closure. And once I had escaped the wet and the humidity, I would be continuing to a fourth expedition. My problem was that I was trying too hard. You cannot control nature.

Once again the jungle fought a war of attrition on both me and my team and we all began to fall apart. One of my teammates, a search and rescue pilot from Cornwall, had lost control of his feet.

We had been in the jungle for ten days. That's ten days of wet feet. Overnight, I had gone around my teammates ensuring they were keeping good hygiene and drying their feet. You cannot overestimate the importance of dry feet on an expedition. They are your wheels. If they're not in good condition, you aren't going to get very far. Foot maintenance can mean the difference between expedition success and failure.

The fear stemmed from something that was known as trench foot during World War II. In the jungle we were swimming and wading for half the day in our jungle boots, which would squelch with water and saturate our socks. The key to a successful expedition was the ability to dry your boots, socks and feet around the fire before bed. If you failed to dry your feet and socks each evening, the result was that damp skin would turn to sores, which eventually turned into open wounds. The drying took time and patience every night. You had to endure the biting insects and the heat of the fire, but without the effort, you risked getting trench foot.

It is hard to describe the pain of trench foot, but it's enough to say that I saw grown men cry in our team. It was so painful and debilitating that we were forced to make a detour to a local village for some medical treatment.

It was at this point that we endured the worst insects of the whole journey. Our succulent British skin was far too tasty for the local mosquitoes and sand flies which, tempted by the exoticism of the full English Breakfast, gorged away on our pale bodies. It was one of these sand fly bites that, unbeknown to me, contained a flesh-eating parasitic bug responsible for a tropical disease known as leishmaniasis, which in its worst form can be lethal to humans. I would end up receiving two courses of chemotherapy at the School of Tropical Diseases to get rid of the disease that kills thousands after my return, which didn't strengthen my bond with the jungle.

So what can we learn from this whole tale? The key to it all was in the attitude. From the beginning, I lost control of my positive mindset and gave in to pessimism. It was like my childhood all over again, in which the anticipation of failure inevitably became reality. That feeling of helplessness turned to resentment, and the wilderness, in her various forms, won, mainly because I had the wrong emotional mindset to cope with her.

Of all the journeys I have been on, the single expedition through Peru was the one in which it felt as though nature was against me rather than with me, and it seemed as though I had turned it into a psychophysical battle of man versus the wild.

It is also a demonstration of the importance of psychological preparation. So many people approach an expedition with the sole goal of physical readiness, but so often that is only the half of it.

After the Canyonlands, it took me ten years to build up the courage to return to the jungle. It had begun to represent the uncontrollable and the fearsome. In my mind, I had exaggerated the discomfort and the oppressiveness of the trees. It had become my 'nightmare' environment where I had no control of the narrative. I used to dread TV proposals that suggested a return to the jungle.

A Glaswegian called Chris Clark lured me back. Chris moved to Brazil in 2004 with his Italian wife. They divorced and he married a local woman called Artemisia. He has become an environmental campaigner, trying to make the area a federal nature reserve to protect the rich flora and fauna. Turtle hunting, overfishing and illegal logging have all threatened this remote, lawless part of the Amazon rainforest and Chris had taken on the task of becoming a custodian to the region. He and Artemisia had a simple stilted house on the banks of the Amazon where they lived off a diet of fish from the river and crops grown on their small allotment.

I visited as part of my Channel 5 series *New Lives in the Wild* in 2018. To get there, I travelled to the Amazon port of Manaus where I joined a boat for the 24-hour journey north, up one of the tributaries towards the Venezuelan border. It was a very different journey to the one I had taken all those years ago as a fresh-faced teenager; this time the boat was smaller, as was the river. We hugged close to the shore. We could hear birdlife and the splashing of pink river dolphins. It was magical, particularly from the relative safety and comfort of our boat.

My heart was pounding with a mix of excitement and trepidation as we headed deeper into the Amazon rainforest. It was the rainy season and we sailed past submerged trees. The dead calm waters reflected a perfect mirror image of each one, giving the impression of double-headed flying trees. It was surreal and beautiful. It felt like a different world.

Three days after leaving the comfort of my home, I arrived at Chris's home in Xixuau. I have rarely felt such fear as I stepped from the comparative safety of that little boat that had been my home, my safe place amid the imposing unfamiliarity of the rainforest all around, and into the heart of the jungle. The heat and the humidity were stifling. I struggled to control my anxieties and the loud inner voice that predicted doom and misery and flesh-eating bugs.

My frame of mind wasn't improved when Chris showed me to my bed, only to discover a huge tarantula crawling across the mosquito net above.

My first real test was when his daughter, Shania, asked if I would join her for a swim in the river. That would be the anaconda and piranha-infested river, in my mind. Usually unfazed, I found myself distinctly unsettled by the prospect of swimming in that river. I watched as she climbed into the dense canopy of an overhanging tree and leapt into the brown waters. She disappeared under the surface for what felt like an eternity. I was convinced she had been snatched by a hungry crocodile.

She giggled as her head broke the surface and she hollered my name, 'Ben! Ben!' Shania was eight, the same age as my daughter. She wasn't scared, so why was I? Perhaps it was a psychological hangover from my last jungle experience, maybe just the unfamiliar surroundings. Perhaps it was just that the jungle wasn't my environment.

I looked at the murky waters and then at the squealing, happy Shania. I clambered up the tree until I was hanging over the river below. I closed my eyes and I leapt. And in that moment, the jungle suddenly became familiar again. The fear had suddenly vanished.

Of all the environments and landscapes that I have known and experienced, the jungle is without doubt the one that has tested me the most. In a world where we want to have full control over everything we do, the jungle is a reminder that sometimes we just have to let go and relinquish control. Like the polar regions, you adapt or you die; you accept the environment and work with it, rather than trying to make it work for you.

While it is impossible to tame the jungle, the jungle had tamed me; my experiences had taught me to let go, but above all, I had learned that without hardship and discomfort, it is impossible to achieve real success.

CHAPTER SIX

THE DESERT

'What makes the desert beautiful is that somewhere it hides a well.'
Antoine de Saint-Exupéry

I love the desert. It is one of the cleanest, simplest, most uncomplicated of environments. That's not to say that the desert is straightforward, but it remains comparatively consistent and it has a great healing quality. The clean air, the unique desert light, the pink sunsets and the red sunrises all create a heady mix of divine spirituality that can move you to your core. Deserts are not easy on the body or mind by any stretch of the imagination, but therein lies their quality. Through hardship comes healing. We have already seen in this book that there is a direct link between a challenging environment and recovery.

There is rarely a desert perfection. Deserts are either too hot or too cold. Too windy or too oppressive. The sweet spot of perfection is often fleeting. Deserts experience tremendous diurnal temperature ranges, which can soar to 45 degrees C during the day and to minus 10 degrees C at night. They are susceptible to blistering sandstorms that strip hair from your legs and they are home to some pretty venomous scorpions, spiders and snakes. What's more, they offer little food or water. On the face of it, they are dead environments. But despite all of that, compared to jungles and oceans and mountains, deserts feel like 'clinical', clean and controllable places.

The desert is synonymous for me with *The Little Prince* and the French Foreign Legion. *Lawrence of Arabia* and the Pyramids. The Paris–Dakar Rally and the Marathon des Sables. Sand and dunes and little else. But scratch beneath the surface and not only will you discover a world of hidden wildlife, of magical oases that spring up like something from a fairy tale, of hallucinogenic mirages and of pristine light, you also have an otherworldly place that demands attention. For me, the desert has always held an allure and a fascination.

For many hundreds of years, the desert has been used as a place for restoring mental well-being and recovery. Even today many of the world's most famous rehabilitation clinics are in desert environments. I know many people who have set off into the desert for their own personal enlightenment because there is a great spirituality that permeates the landscape. Perhaps theology has a part to play – after all, the Three Wise Men arrived in Bethlehem, having trekked through the desert on their camels.

In my innocence, I always thought the sand dunes that I found next to the beach would be similar to the ones in the 'real' desert, until I had my first true experience and realised how very wrong

I had been. It is important to understand that there are different kinds of desert: hot and dry, semi-arid, coastal, and, yes, cold deserts.

My first encounter with the desert was pretty extreme. It was 2007 and I was in a pub with friends; we somehow got onto the subject of endurance running, and one of them dared me to run the London Marathon. Now I would be the first to call myself naturally unsporty – but I am also not one to shy away from a challenge. So I suggested that not only would I run a marathon, I would run SIX of them, back to back, and enter the gruelling Marathon des Sables, the infamous race that takes place every April across 150 miles of the arid expanse of North Africa's Sahara desert. I was very much a sporting novice (I hadn't yet rowed the Atlantic) and this would be my first big adventure race.

In the past, runners had been lost and had even died taking part, but none of that put me off, not even the fact I had never run in my life. At school I was allergic to sport. I was the kid who always had the sick note to get out of PE. And now, I had committed myself to one of the toughest races on earth.

Nervously, I called the race organisers for a place. They apologised that the event was already heavily over-subscribed, but said they would put my name on a waiting list. There were, they informed me, a thousand people waiting ahead of me.

I relaxed. I had honoured my commitment, but I wouldn't have to race the darn thing. Then, just eight weeks before the race, I got a call from an unknown number: 'Congratulations,' chirped the voice on the phone, 'a thousand people have dropped out.'

And in that instant I found myself with a place in a 150-mile multi-marathon across the Sahara with just eight weeks to prepare.

Not one to panic, I joined my local gym, bought some brand new running clothes and found a fitness trainer. With tags still hanging from my shorts and trainers, I asked him if he could help get me ready for a little race in a few weeks' time.

'A 10k run?' he asked enthusiastically.

I nervously shook my head. There was a sharp intake of breath.

'A half-marathon?' he asked.

When I revealed the magnitude of the race, he refused to train me in case I died and my parents sued for negligence. So I trained myself, getting up each morning and running for as many hours as I could.

Eight weeks later, I arrived in the remote town of Ouarzazate in southern Morocco, along with 700 elite endurance runners from all around the world.

The only way to describe that week is 'painful'. In fact, the word 'pain' doesn't really do it justice. I have rarely experienced pain like that. The sand would cascade into my shoes and embed tiny particles into my socks, and the friction of sock against skin would act like sandpaper and rub layer upon layer of skin from my poor feet. Then blisters would form on the raw skin. At the end of each day, the team of French doctors would set up a mobile clinic called 'Doctor Trotters' from which they would treat our battle-weary feet.

To avoid infection, the doctors would use a scalpel to cut the skin from the blister, then drain and douse it with iodine; the next day, the bandaged and heat-swollen feet would rub still further, creating more blisters and abscesses. I mean it when I say I have rarely experienced such agony. To date, the MdS remains the most painful experience of my life. The only way to get through it was to try and ignore it. And despite the pain and the misery of those endless miles of dunes and sand on my tattered, blistered feet, the desert was a great healer.

In my suffering, the desert took on a restorative quality. I felt a surge of power unlike anything I have ever experienced elsewhere. It was partly environmental but also the camaraderie of hundreds of strangers all enduring and suffering together. Each day was like a rolling wave of emotions that would begin with utter wretchedness each morning when I woke up, unable to walk on my tattered, blistered, bloodied feet. I would shuffle around camp. Just walking for a pee was agony. Incidentally, I wasn't the only person who struggled with morning ablutions. On the first day, when everyone had fresh feet, people would walk 100 metres from camp to pee and poo, but by the last day, people were defecating on their own doorsteps, quite literally.

I would begin the day hobbling and wincing in pain. As we gathered on the start line, the idea of a whole marathon ahead of us seemed as dreadful as it would be miserable. But then, slowly at first, the adrenaline would begin to surge around my crippled body and incredibly, as the start gun sounded, my body lunged forward to propel me across the desert. The first few steps felt like running with shards of shattered glass in my shoes, and then, slowly, the pain would lift, as did my mood.

Pain and misery slowly gave way to elation and wonder at the power of the mind. Then an hour later, the high would wear off, probably at the same time as the painkillers, and the pain and misery would once again loom large, clouding my spirits with a big black cloud of impossibility. But here is where the desert began to work her magic. The arid beauty of the landscape would seep into my body, lifting my spirits and giving me a mental high. Like a rolling ocean, each day would be a waxing and then waning tide of emotions that would sweep me up and down like ocean flotsam and jetsam. I was a willing slave to the desert.

The deserts of the world have long been harnessed for their therapeutic qualities. Palm Springs in California is the capital for drugs and alcohol rehab, while the Sahara has long been used to heal physically and mentally, but here under the blistering Saharan sun, despite the suffering, the pain, the misery and the deprivations, I had an epiphany. It was such a profound feeling of clarity and direction.

Don't get me wrong, the race wasn't enjoyable but it became deeply spiritual. The desert has an invisible power to provide clarity and vision. I'm not sure if it is the dry, clear air or the heat, or the sand, but it has a cleansing quality that I have rarely experienced elsewhere. It felt like a contentment through my connectedness with the landscape. Like a spirit animal, I think we all have a landscape that suits us. An environment that matches our own inner feelings of complacency or comfort. A sense of belonging.

For the first time in the race, I had used mind over matter. Despite the pain in my feet and the exhaustion in my lungs, I was able to summon an inner spirit to help me and guide me. Until then, I had still been controlled by the powerful inner voice of doubt, but for the first time I was able to control my own mind and rewrite the narrative. I felt braver, stronger, and I began to believe that anything was possible with conviction and determination.

The MdS was the first time I had really surprised my critics. I returned walking taller, although technically I was actually shorter, bent double over the crutches that I required for my damaged feet.

The desert had been my guide and my companion. It had been illuminating and it had helped clear my cluttered mind. I was able to see and think more clearly. I sometimes experience a sense of burden and chaos within. It almost always happens when there is clutter, both physical and metaphorical. If I am overwhelmed with stuff both in the house and in my mind, it leans on me like a heavy cloud. I am unable to think straight until I have tidied. The desert, with her Zen-like tidiness and lack of clutter, offers the perfect escape from the chaos and jumble of life.

My first visit to the Sahara desert of Libya was unforgettable. I sat with my face pressed to the plane window as we flew over hundreds of miles of yellow sand dunes below, tears streaming down my face. Fear and despondency overwhelmed me.

As I stared at the seemingly endless desert below, I felt hopeless. A failure. Clutched in my hands, a crumpled copy of the *Sun* newspaper in which a tear-stained article headlined 'Fogle is a Dog's Dinner' concluded that not only was I too ugly to be on television but I was too posh and also a terrible presenter.

I had just concluded my first live BBC prime-time programme as host of Crufts, the world's biggest dog show. For three consecutive nights, I had anchored the show for two hours. It was my first time presenting on live television and my first time using an autocue. Or trying to – I am so short-sighted that I couldn't read it clearly, so I squinted and bluffed my way through. Viewing figures went up but the 'professional' dog show community didn't like my style, which was more mongrel than pedigree.

The resulting news article had come at a time of immense emotional vulnerability for me. My career was still in its infancy and I struggled with its fragility. I was still lacking in confidence and the article seemed to confirm all my fears. But the show, as they say, must go on and ahead of me was an expedition across the vast, arid dunes of the Libyan desert.

It was there that I experienced my first desert storm. Both terrifying and wonderful, a sandstorm is unlike any other. The sand gets whipped into a ferocious blizzard that obscures everything. For two long days, the violent storm howled around us. We used ski goggles to protect our eyes, and the film crew who had been following the journey were forced to join us and use the vehicles to provide a barrier against the relentless wind. We used our shemagh scarfs to wrap around our noses and mouths to save our lungs from the powdery thin dust swirling around. Any exposed skin felt like it was being scrubbed with sandpaper.

For two days, we huddled next to the vehicles, our backs to the wind, lost in our own misery. And misery it was; there is little excitement or relief in a sandstorm. When the winds finally abated, everything was covered in several inches of sand and dust.

The desert took on a whole new quality afterwards: what had previously appeared harsh and barren suddenly looked friendly and serene. Like a cleansing, the silence and calm after a desert storm was extraordinary.

That sandstorm erased the news story from my mind and brought me crashing back to the present. It was a reminder that you can't live your life always looking back. Sometimes you need to accept your plight and move forward.

That desert storm might have stung but it helped make me more thick-skinned. It gave me a context. When I think back to my reaction to that article now, it is of course laughable. I have experienced far worse in the intervening years and have become much more impervious to criticism from my time in the wilderness.

The desert has also been a place where I have experienced near-fatal disaster. It happened when I was on a journey across the coastal Atacama desert of Chile, leading an expedition for the TV series *Extreme Dreams*. This was the last part of the 12-week journey that had started with the failed climb of Chimborazo in Ecuador, as described in the previous chapter.

For three long months I had sweated and toiled my way across South America with a rotating film crew and a new cast of team-mates for each adventure. We had climbed across the Andes, trekked through the jungle and down into the Canyonlands of Peru, and now I was on the final leg, a ten-day trek across the Atacama desert.

As far as I was concerned, we had left the best and easiest until last. I felt I knew the desert, and it would be a fittingly healing conclusion to the South American odyssey and a reprieve after weeks in the jungle. Little did I realise that it would turn into one of the hardest and most dangerous journeys of my life.

We met in the 'Ibiza of the desert', San Pedro de Atacama, a Chilean town that draws hippies and trust fund millionaires from around the world to its picture-postcard adobe houses and boho fresh-juice cafés. For the first week we trekked across the barren landscape, across salt desert and rocky wasteland, until we slowly began to ascend. Our goal was the highest point in the desert at 4,000 metres. Surprisingly high for a desert environment, but nothing compared to the 6,300 metres of Chimborazo in Ecuador.

Although we prepared ourselves for breathlessness, altitude was never our main concern. We were more focused on the heat and the aridity of this harsh landscape. As soon as we reached 3,000 metres though, things began to unravel.

One of my teammates, Dave, a search and rescue helicopter pilot, began to complain of headaches and chest pains. His condition began to deteriorate and our expedition medic, Alex, took the decision to place him in our portable hypobaric chamber. The size of a large sleeping bag and shaped like a giant Smarties tube, the pressurised chamber is operated by a foot-controlled vacuum pump as a way of reducing the effect of high altitude by rapidly increasing the concentration of oxygen in the blood.

We placed Dave in the chamber and pumped it up. A small plastic window allowed us to check on his condition. It certainly wasn't ideal for someone who suffered from claustrophobia, but this was not the time to worry about that; we had to save his fast-deteriorating health. We took turns with the foot pump to maintain the pressure in the chamber and ensure a steady stream of oxygen to his blood. In less than an hour, his condition went from critical to stable.

We called our medical team who told us that there were no helicopters in the region and that the only means of evacuation was by road ambulance, which would take a day or more to reach us. We would need to stay put and, more worryingly for Dave, he would need to stay put inside the chamber. Now feeling much better, he begged to be released from his little plastic coffin. We had to cup our hands over the little plastic see-through window to communicate.

It was heartbreaking to hear his pleading to be released. Panic-stricken, he reached into his pocket for his penknife and threatened to cut himself out unless we released him. But if he pierced the chamber and depressurised it, then he would likely die. We pleaded with him to stay inside. We took turns. One of us worked the foot pump continuously while another would keep him company through the little window. For more than 24 hours we pumped and pleaded with him, until finally a 4x4 ambulance, sirens blazing, appeared on the horizon, the medics arrived and treated him before whisking him off to the closest hospital.

It had been a shock, but we regrouped, secure in the knowledge that he was in good hands, and we continued our journey across the Atacama desert. Just a day later, and only a little higher in altitude at 3,500 metres, the same thing happened to our director, and he too was medevacked off the mountain.

Three days later we finally arrived at our goal, a small peak in the middle of the vast desert. We were alone for hundreds of miles and I had completed my 12-week journey across Latin America. There were tears and elation and relief that we had reached the end. As we made our way to the waiting minibus, Alex, our medic, who was also a former army officer, screamed out loud, 'Stop! Nobody move!'

Shocked and confused, we all froze. We looked around. There was nothing visible to concern us. Was this the desert playing tricks on us again?

'Mines,' he explained. To our horror, we had somehow strayed into a live minefield. He had spotted the tell-tale sign of a tiny metal trigger sticking out from the rocky desert floor.

This region had been the scene of a long war between Chile and Argentina, and it turns out the desert had been booby-trapped with thousands of mines, many of which had never been cleared, and we had now strayed unwittingly into the middle of it all.

With the utmost care, Alex placed a handkerchief on the ground near to the mine and he instructed us to step carefully around it and onto the minibus. It was a close call. I have never been so relieved to get out of a place in my life.

We were soon back in San Pedro de Atacama and reunited with the rest of the team who had been suffering from altitude sickness. We spent several days celebrating and recovering until it was finally time to return home. We boarded the plane in Santiago for the long journey east, back to Europe.

I was finally going home. We had made it. I watched from the flight window as the desert sprawled out below us. It would soon be a distant memory. But the desert wasn't quite done with us yet.

An hour into the flight, the Captain made an announcement. 'Do we have a doctor on board?' My heart sank. I soon discovered that one of our team had relapsed. They were already on supplementary oxygen but there wasn't enough to last us the journey to Spain. I peered out of the window at the vapour trail as our plane dumped her fuel, way above Brazil, before making an emergency landing into São Paulo. Ambulances raced down the runway to meet us, sirens and lights blazing.

Like I say, always expect the unexpected from the desert.

That's not to say the desert can't be a comfortable place to live. Over the years I have spent time with those who have made the desert their home, but it was one person in particular who had transformed the desert not so much into an oasis as a utopia. Nearly a decade after running across the Sahara in the Marathon des Sables, I returned to Ouarzazate in Morocco to visit Australian Karen Hadfield.

Karen had set up an artist's retreat called Café Tissardmine in the middle of the Sahara; an oasis of her own among the towering sand dunes all around. A 12-hour off-road journey from Marrakech, within sight of the Algerian border, she had built an extraordinary home of sand. To call it a home doesn't really do it justice; built from the desert sand was her fortress, her riad.

This riad consisted of one main building in the centre of a large sprawling courtyard filled with plants and flowers. Around the edges were the khaimas, the Berber tents, which had been made from dozens of heavy carpets used to create rooms. The effect is as magical as it is homely and comfortable.

The sand absorbs all noise and the result is an extraordinary silence that is enhanced by the light and the clean air. I have rarely felt as comfortable as I did at Café Tissardmine with Karen, who walked around in an exquisite Moroccan cloak known as a djellaba.

Karen had moved to the desert after tiring of working in festivals in Australia. She wanted to escape and the desert was her calling. She visited Morocco on holiday, fell in love with the country, the people and the desert, and a few months later she returned to build a retreat. She met Berber Youssef and together they realised their dream of building a home in the desert.

Shortly before I visited Karen, Youssef was tragically killed in a car accident. Karen was heartbroken and still mourning his loss, but the desert had been her spiritual healer and she had resolved to remain and continue the dream they had started together. Spending time with Karen opened my eyes to seeing the desert as both a home and a spiritual retreat. I could see how the impact of her loss had been softened by the divine healing qualities of her surroundings.

Shortly after I left Karen, her dog, Diggedy, escaped and joined the famous Marathon des Sables as it passed the riad. Re-named 'Cactus' by the runners, Diggedy made headlines all around the world as he became the most famous living thing ever to run the MdS and unwittingly became a global superstar.

I feel sure that Diggedy overheard some of my own tales of running the MdS a decade earlier and felt inspired to run it himself. He ended up with a book deal and even a Hollywood film, which was a little more than the pair of crutches I ended up with after my own efforts.

But it was a desert journey with my friend James Cracknell in 2011 that would really test me and remind me that the desert isn't always so hospitable.

* * *

After rowing across the Atlantic and walking across Antarctica to the South Pole, James and I had decided to conclude our adventure friendship with one final expedition. We settled on a mountain bike challenge across North America: a 3,000-mile off-road mountain bike trail from Canada to Mexico down the entire spine of the Rocky Mountains. It would give us a chance to test ourselves, our bodies and our friendship one last time.

For a year we cycled all over the world in preparation for this massive undertaking, but just a week before we were due to start the race, James was in a bicycle accident. He was cycling along a road when a petrol tanker crashed into him, catapulting him at 80 mph from his bike onto the tarmac. His helmet sheared in half and he sustained a life-threatening, life-altering head injury. He was in intensive care for weeks. In hospital for months and recovering for years; in fact, I think it's fair to say he is probably still in recovery today, nearly a decade later.

Aside from the physical injuries, it is impossible to overestimate the psychological effect of a major head injury on a victim as well as on their loved ones. The injured person experiences a change in personality, and the loved ones have to live with the consequences. James had to learn how to do things again: he had to relearn cognitive thinking; he lost his sense of taste; he lost empathy, and more importantly he lost the filter in his brain that would normally stop him saying inappropriate or offensive things. The consequences for James, his family and his friends, were devastating.

For a couple of years, he rebuilt his shattered life, and one day we decided it was time to complete our final expedition. We agreed not to attempt the mountain-bike journey but instead we would head into the desert for physical and mental healing.

Of course, this was not going to be a five-star Nevada desert retreat. We would experience the desert through adversity, not by running 150 miles across it, but instead by trekking 750 miles, recreating the feat of a famous British desert pioneer and explorer called Wilfred Thesiger.

'In the desert I had found a freedom unattainable in civilisation; a life unhampered by possessions, since everything that was not a necessity was an encumbrance,' said Thesiger in his classic adventure tale, *Arabian Sands*. It is a simple, astute observation and one that was one of the big draws for me to return to the desert with James.

The Empty Quarter, or the Rub' al Khali, is a vast desert expanse covering 250,000 square miles of the southern Arabian peninsula. In a region defined by deserts, the Empty Quarter has a reputation for being one of the most inhospitable and daunting of all. Down the centuries it has tested the forbearance of kings, nomads and adventurers. As for Thesiger, after serving in the North African campaign in World War II, he crossed the Empty Quarter twice, in 1946 and 1947, doing his research from the back of a camel.

One of the driest places on earth – even drier than Death Valley in California – the Rub' al Khali has attracted traders and nomads throughout the region's rich history, and there are tantalising tales of a lost city hidden within it. Known variously as the 'Atlantis of the Sand', Ubar and Iram, many adventurers and treasure hunters have searched for it, but no one has ever found it.

James and I weren't looking for a lost city; instead we were looking for our lost friendship. Our relationship had been strained since his head injury. Our lives had gone in different directions. We both had growing family commitments and the added pressure of his head injury created tensions that had caused a rift. In some ways he had become a stranger again. I didn't recognise him as the man with whom I had rowed the Atlantic and crossed Antarctica.

By recreating the crossing of the Rub' al Khali together, we had the opportunity to get to know and understand one another again, without the background noise of society. Free from distraction and, as Thesiger had observed, 'free from possession', we were free to rebuild our friendship. We would use our journey not just as a physical endeavour, but as a balm for our mental welfare.

By now I had become a father. Ludo and Iona had changed my life. They had also changed my attitude when it came to risk. Until I became a father, risk had been key to rebuilding that shattered childhood confidence, but with parenthood came responsibility. Almost overnight, I became a sensible dad. Not only did I start wearing dad clothes, doing dad dancing and telling dad jokes, but I became risk averse. I wanted to conform to society's expectations of what dads do, and dads don't go rowing across oceans or swimming with crocodiles. Sensible dads protect and care and provide and nurture.

And yet ... although becoming a father had changed me greatly, deep down I still craved adventure. I wanted to feel young and free again. I wanted the excitement of the adventure; the thrill of an expedition; and the sentiment and nostalgia of my friendship with James again.

It had been five years since James and I had crossed the Antarctic, and the BBC wanted a return adventure. They liked the idea of a recreation rather than a race. They were curious as to how we would cope under different conditions.

Following in Thesiger's 1946 footsteps, our journey would take us from Salalah on the south coast of Oman, due north to Saudi Arabia where we would then follow the frontier east for 250 miles. We would do the whole thing old style without any modern tools: no cars, satnav, dehydrated foods nor desert boots. Instead we would wear local dish-dash tunics and sandals, rely on maps and a compass for navigation, eat dates and dried fish and use camels for transport.

There would be no race but we had implemented a series of guiding rules to ensure authenticity. When I say 'we', I use the royal 'we', meaning the production company. It is probably important to explain that where the Atlantic row and the polar trek had both been 'ours', James and I had lost control of the Empty Quarter quite early. We had lost it because of laziness. The production company had offered to do the legwork and the planning and we had happily deferred. The result was that we lost ownership, which is a pretty important part of any expedition.

As many people know, it is often the planning and the preparation that are more fun than the journey or expedition itself. But with the Empty Quarter we fast-tracked from idea to expedition without the planning. We left that to the production company, and as a result, the adventure never quite felt like our own.

Neither James nor I had ever worked with animals on expeditions before. I had limited experience with camels from Longleat Safari Park when filming the BBC's *Animal Park*, but there, my experience had been largely limited to feeding them occasionally.

Working with camels in the desert was a whole different kettle of dried fish.

While the Atlantic and Antarctica had both tested our fortitude and resilience, the Empty Quarter would test us with unfamiliarity. Stripped of modern-day kit, we were forced to learn everything from scratch: from making a fire, to cooking, navigating and animal husbandry. But first we needed to learn how to survive in the desert, and more importantly, learn how to work with camels. It was perhaps this little detail that proved to be our weakness.

Although we both love animals, we found that camels lived up to their reputation as notoriously difficult to work with. It can take years to build a relationship and a rapport with a camel, but James and I didn't have the luxury of time. We had just a week to select, train and prepare the camels that would become our teammates for the duration of the crossing.

We had been recommended a local guide, Masallem, who would help us assemble all of the original equipment, teach us the basic skills to stay alive and help us with the camels. But first we had to buy them.

I will never forget the day we went to the camel market to buy our four camels, two each. A scrubland in the desert had become an inland port for hundreds of these 'ships of the desert'. I could smell the pungent camels before I could hear them. A cacophony of groaning, grunting and baying that floated across the otherwise silent sands.

The camel market was a hive of activity as buyers and sellers haggled and bartered. It was comedic to watch camels squashed into the back of pickup trucks and even taxis, their legs bound together at the knees to turn them into little parcels. It was amazing to see what a tiny space a camel could be squashed into. They looked surprisingly chilled and relaxed as they chewed the cud.

We were late to the market. It was a little like being the last to the fruit seller. The best fruits had been sold, and all that remained was the overripe, bruised stuff, not that we realised this at the time. After all, one camel looks and behaves very much like another to the untrained eye.

We pointed to four camels. They spat angrily as Masallem patted their moulting flanks and looked at their teeth, which looked worse than Shane MacGowan's. And after a brief negotiation, they were ours. It all seemed very rushed and unscientific. I have spent more time selecting wallpaper for the house than we did those four camels, and their lives relied on us, and ours on theirs. And wallpaper could be papered over if you made a mistake.

We then spent several days in camel school learning how to look after our new teammates: what to feed them, how to ride them and how to pack them. Sounds easy? It isn't, especially when you are working with 50-year-old equipment that is falling apart at the seams.

Travelling alone, we would need to carry food, water, bedding and camel feed. Since we had decided to recreate the expedition as closely as possible to the original journey that Thesiger had taken, this included no underwear, and just dried fish and dates for food. The camels too would eat our dates. It sounded straightforward but, as is often the case, the theory was far easier than the practice.

We set off under the scorching sun, and after just a day our bodies began to fall apart as we endured one of the wildest places on earth. Dehydration, heat stroke and hunger all began to take their toll as the sand rubbed at our sandalled feet. I genuinely don't think I have ever been to a landscape as otherworldly and wild as the Empty Quarter. On the face of it, it is an endless, lifeless desert of soaring dunes and sand, but it is this endless magnitude that creates a sense of wonder. It really feels like nowhere else.

The six of us, James, myself and our four camels, marched across the scorching hot sand. As we crested the brow of the first dune, to reveal more red rolling dunes as far as the eye could see, the view took my breath away. I think it might be one of the most astonishing places I have ever seen and that is saying something. The desert is wild and harsh but it is not threatening, or at least I didn't think it was threatening at that time. It sometimes looks impossible and it is confusing and exhausting to navigate, but it has a pristine cleanliness about it that is difficult to find elsewhere. Perhaps it is this cleanliness that creates a sterile, pure, non-threatening environment. It is hot and arid, but there is a simplicity to its wilds.

Each evening we would sleep on the sand and allow the camels to roam. We had been advised to 'handcuff' their legs to stop them roaming too far, but were too soft and left them untied, which meant that each morning we would spend the first few hours scouring the desert for our free-range camels. Exasperated with this routine, we decided to tether one of their legs to ours as we slept. I woke up being dragged across the desert by a wandering camel.

As an animal lover, I stressed and worried about the welfare of our camels. We only had limited water, which we needed to replenish at natural springs along the way, and the camels refused to eat the dates we had packed to feed them. We were out of our depth, but more worryingly, our hearts and our heads weren't in it. For a week we got lost, we got stuck, we got the camels stuck and we got lost again. We fought and argued. James became irritable and I lost my patience. I have mentioned that one of the effects of his head injury had been the loss of his filter, which would normally control what came out of his mouth. James was already prone to rudeness before the head injury; now, without the 'good-manners' filter, it was like the volume had been turned up. We were failing to bond with the camels and we were struggling to bond with one another.

One morning, while trying to entice the camels with a handful of dates, one of them mistook my hand for a far tastier snack and bit into the palm, its teeth cutting to the bone. It was a deep open wound that required stitches. Another psychological blow. I've still got the scar as a souvenir. We were making slow progress and to make up time we decided to give our sandalled feet a rest and ride the camels. If you have never ridden a camel, let me give you some advice. Don't. They don't like it and you won't enjoy it. Using tiny wooden saddles from the turn of the last century, our thighs rubbed uncomfortably as the camels slowly marched across the desert. Without someone to lead them, the camels slowed to a saunter that was slower than our walking speed.

We tried to gallop. Big mistake. The camel I was riding galloped, while my second camel, her harness tethered to my saddle, decided to sit down. Yes, she actually did that mid-gallop. Let me tell you something. A camel with a low centre of gravity has the power to stop a galloping camel in its tracks, rip my saddle from its back and send me flying through the air.

Did I mention that camels are quite tall? Especially when you have been thrown from your saddle. I landed hard on my side, the air wheezing from my body. Winded, I struggled for breath, but it was a sharp shooting pain in my chest that was excruciating. I had broken my ribs.

I said our hearts weren't in it. This felt like the final straw. Negativity and pessimism clung to us as we walked in silence.

I regret the timing of that journey. If James and I were being honest, we never really wanted it. It was never 'our' journey. Laziness and indifference in the planning stages had meant we were never truly invested. It was like we were passengers on someone else's journey. The Empty Quarter was more about the television show than the friendship, and the result was a gruelling schedule that negated the healing, powerful force of the desert. I couldn't help comparing it to the Marathon des Sables, which also had a time pressure, but that had unified the hundreds of runners rather than dividing them.

James still wasn't comfortable in his own company. But he resented the fact we had to have a medic with us, accompanying the film crew in their vehicles, for insurance purposes. A dark cloud descended. This negativity was like a toxic shroud that clung to us in the middle of that beautiful place. One day I even had to break up a fight between James and a camel. I kid you not. He was about to punch it and I had to stand between the two of them.

By now I was in agony. Unsure as to the severity of my rib breakage, we were medevacked by the film crew to a hospital. Thanks to a heavy dose of Valium, we bounced across the desert for eight hours until we reached the nearest town. I had broken several ribs and we decided to abandon the challenge. It was a humiliating end to the expedition. If I'm honest, and I hate to admit this, I was relieved. I wasn't enjoying it at all. Despite a love of the desert, it had got under my skin and I had lost control. We returned home, much to my family's surprise, and for a month I convalesced.

Five weeks later we returned to finish what we had started. Neither of us is a quitter and we weren't about to give up on our final expedition together. We had successfully rowed the Atlantic and crossed Antarctica; we weren't going to be beaten by the desert. We rejoined Masallem, who had been looking after our camels. (Although I have always been suspicious as to whether they really were the same ones. They looked, smelled and acted differently, not that we ever found out. As I've said, one camel looks very like another.) We repacked and restarted our desert journey.

Summer had broken and the sun beat down with an even more ferocious intensity, but our minds were in a better space. We'd had time to digest our previous experience and understood what it was we were doing, and both of us felt more invested. It had been our choice to return, and with that simple decision, the power had returned to us.

In the first phase of the project, it felt like we were the actors in somebody else's play. It was as if we had subliminally checked out. The Atlantic had been *our* expedition. *Our* challenge. *Our* planning. It was the same with the South Pole. We would stand or fall by our decisions and any mistakes were our own. We couldn't blame anyone but ourselves. But here in the desert, there was never a sense of ownership, until now. Suddenly it felt like *ours* and we took the bull by the horns (or the camel by the hump) and ran with it (or ambled as camels do). This time there was no fear of the unknown.

The desert became less of a burden. She was hotter and harder, but there was a renewed energy that reminded me of my previous desert encounters. That early race across the Sahara in the MdS had been so empowering and healing, and now here, in another great desert, it was nice again to be experiencing the therapy of the desert.

My friendship with James blossomed and our working relationship with the camels improved. And three weeks later we completed the journey. It was a very different experience to anything we had done before, but for James and me, in a strange way, I think it was one of the most meaningful of all our expeditions. With the help of the healing qualities of that sterile, arid environment, James and I were able to rekindle our friendship. I think for both of us, it was closure. It was some sort of punctuation mark at the end of our adventuring partnership. Don't worry, that doesn't mean it's the end, but it was a fitting finale for the tale of two compete strangers who took on the ocean, the ice and the desert, and came out the other end.

Ours were never wins. They were not battles. They were encounters. I like to think that we had shown respect and reverence, and in return we hadn't won, but we also hadn't failed.

Something extraordinary happens when you set foot in a desert. Nature takes over. The desert, with its sterile environment, is a silent assault on the senses. In a world with an alarming growth in mental health problems, the desert is like a vaccine against the pressures of modern society. It is a place of great spirituality: a magical, otherworldly landscape that can build spirit and hope and silence doubt.

CHAPTER SEVEN

WILD FOLK

'In every walk with nature, one receives
far more than he seeks.'
John Muir

My paddle dipped into the thick, brackish waters. Dense, green vegetation enveloped me like a cloak as I rowed gently upstream, against the invisible flow. A mighty clap of thunder preceded a biblical downpour that threatened to fill my entire canoe. In seconds I was soaked to the core. My clothes clung to me like a heavy second skin. Through the torrential rain, I weaved between submerged trees and branches. The rain fell so hard on the water that it bounced up, creating the illusion of thousands of balls of water dancing on the river surface. The rain burst soon subsided and was replaced by a hot sun that beat down with such ferocity that the rain turned to vapour, rising in a fine mist that shrouded the landscape.

Between two submerged trees, a huge spider had spun a mighty web. It was vast. As my eyes began to adapt to the jungle-like vegetation, I noticed more webs and spiders. Now was probably not the moment to admit a weakness for spiders. Okay, I'm scared of them. And I had just realised they were everywhere.

I was in the middle of a hot, stinky swamp in southern Georgia in the United States on my way to meet someone who had made the wilderness their home.

When I say 'the wilderness', what do I actually mean? The *Cambridge English Dictionary* defines it as 'an area of land that has not been used to grow crops, or has towns and roads built on it, especially because it is difficult to live in as a result of its extremely cold or hot weather or bad earth'. So what happens to a wilderness once it is inhabited by man? Is an island still an island when it has a bridge? Does a wilderness lose its wildness once it has been tamed or harnessed? Or does it re-wild us. I am fascinated by the notion of re-wilding, whereby we try to return nature to its natural state. But what about us? Is it possible for mankind to be re-wilded? And returned to our natural state?

I have already dwelt on examples of the wilderness and her power to heal and build strength, but what about the simplicity that comes with basic living? Most of us have lost our ability to be self-reliant. The long period of international lockdown after Covid-19 has demonstrated our reliance on others, but what happens when you forfeit dependency culture for self-reliant resourcefulness?

Over the years I have spent a great deal of time 'playing' in the wilderness. It has been my gym, my classroom and my playground. It has been a place of temporary immersion (if you call a year temporary, which I do) – but what about those who have made it their home? Not the Amazonian tribes or the native indigenous populations, but people like me who had grown up in the city.

All around the world, there are individuals and communities who have escaped into the wilderness to try and build their own lives away from civilisation and humanity. These 'wild folk' have decided to embrace the heat or the cold or the drought or the excessive moisture to build themselves a small, low-impact life in the wild.

Most of my experiences in the wilderness have been fleeting. Temporary encounters of short-term immersion that leave little or no trace. But what about those who have developed a co-dependency on the wilderness? What about those who don't just dip a toe into the wilderness, but move themselves entirely? Mankind has a heavy foot. We don't have a great track record when it comes to our relationship with nature. We tend to be takers rather than givers. But there are plenty of examples where people have developed a more sympathetic relationship with the natural world.

For nearly a decade now, I have been visiting people who have abandoned their previous lives for a new one, off-grid, for a television series.

The premise of the programme was simple: to meet the misfits, the wackos and the wild folk who grew beards and dropped out of society to become the madmen in the woods. That's really how the series began, with its working title (still used internationally) *Where the Wild Men Are*.

The assumption was that it would be a light-hearted journey to meet eccentrics, but as the series moved on, we soon discovered that not only were these individuals far more sophisticated and wise than we had first assumed, but also that many of them had breasts instead of beards: women, we would soon discover, can be far more successful living off-grid in the wilderness than men.

If I'm really honest, I didn't think the series would work when it was first proposed to me, partly on account of the month-long 'immersion' suggested by the production company which would in practice require months away from home. And partly because I wasn't sure that there was a big enough audience for what was still a niche idea. My thought was that we would be asking them to watch people chop wood and hunt animals.

The first person I ever visited was John Wells, a former fashion photographer from New York who, tiring of the fashion world, had bought a small patch of desert in Texas, close to the town of Terlingua on the Mexican border.

Here he had created a new life that was part survival, part Mad Max and part art installation. I suppose it isn't surprising that a former photographer had brought his creativity into the wilds of the desert, but at this stage I was still anticipating a gnarly old man with buck teeth and out-of-control facial hair.

John fitted one of those clichés with his beard, but beyond that, he was a long way from the type of character we had anticipated. With his hat on he looked more like a band member from ZZ Top.

He had retreated to a small, remote, unremarkable patch of desert and scrubland close to Big Bend National Park, an area renowned for its beauty and its odd characters. This was where people came to drop off the face of the map. If you were escaping something then this was where you would hide. No one asked questions and there was plenty of space to conceal yourself. (Shortly before I arrived, one of America's most wanted villains was caught hiding out on nearby land.)

John wasn't here to hide, but he was retreating from the commercial world. He had built himself a cabin, a tiny cabin, just big enough for a single bed and a desk. He had equipped it with a homemade 'swamp air conditioner', essentially a metal grill over which water slowly dripped, with a fan to blow air through it. It was rudimentary but highly effective in the 100 degrees C desert heat. John had also built himself a polytunnel in which he grew some crops and held water from the sporadic rainfall.

He had a pet chicken and a pet long-horned cow. He used his creativity to turn his life into an experimental art installation. It was beautifully eccentric, with a broken payphone in the middle of his compound. His little tin house was picture-postcard perfect. Everything had been painted and decorated to resemble art as well as being functional. A Perspex box was used as a shower with a solar heater on its roof, and an assortment of animals added texture. Everything had been considered. The small life he had created for himself was far more than just sustaining and surviving; his life was about thriving and living. Really living.

His life was simple and, on the face of it, pretty ideal. There in the desert surrounded by a cast of odd characters and end-of-the-world-preppers, he had built himself a small utopia. While I was with him, we caught rattlesnakes, wrangled cows with his cowboy neighbours, mucked around with guns and I got to wear cowboy boots. I was captivated.

We visited the house of a nearby 'prepper', who had filled his basement with three years' worth of tinned food, ketchup, water and fuel. He had more than 30 cameras around his house and his land and he would spend his days staring at the 'control centre' – dozens of screens – waiting for the zombie apocalypse.

It was amazing. Mind-blowing. Until now, the wilderness had largely been a place of natural beauty, home to indigenous people. I had previously experienced it on different levels, by visiting it in its natural state and by endurance through expeditions, but this was the first time I was experiencing it through the eyes of an 'alien', as in someone who doesn't naturally belong. I suppose part of the attraction was the fact that it reminded me of my own *Castaway* experience, when we were the 'foreign objects', out of our natural environment and forced to adapt to a new one.

Spending time with John Wells proved to me that the wilderness can be a wild, fun, irreverent place. That desert was not particularly hospitable: it was dry and hot and lonely and, on the face of it, lifeless. The visit shifted my thinking about the wilderness: I became fixated on the possibilities of it, but most importantly, the possibility that it could heal damaged goods.

John wasn't damaged per se, but like many of us, he had become disillusioned with life. He had reached a moment where he asked himself: what's the point? Where is the journey? The difference between you, me and him, was that he did something about it. He walked away from his old life and embraced a new one.

My wild men journey has taken me all over the world to meet strange and interesting types, but what fascinated me was meeting people like you and me. Everyday folk who had given it all up for a new life in the wild.

Back in the mosquito- and spider-infested swamp in the US state of Georgia, I was ready for another confession: swamps are not really my thing. I know there are plenty of people who can see past the stinky water, the catfish, the alligators and the mosquitoes to appreciate the swamp's inner beauty, but I found it difficult to see beyond the palm-sized spiders that hung from every tree.

There I was in a canoe, not far from the city limits and urban sprawl, soaking wet, paddling between spiders' webs and submerged trees in search of my swamp-dwelling wild man called Colbert Sturgeon. He had somehow, against the odds, built himself his own childhood utopia in the swamplands. In that south Georgia swamp, Colbert had created an oasis away from the pressures of the material world. He set lines to catch catfish and he ate roadkill. He not only survived but he thrived. In a wilderness largely overlooked by the city folk beyond, Colbert had spent 20 years slowly gathering in his canoe the materials to build his perfect cabin in the woods, or swamps in this case. Single-handedly he hauled mighty timbers through the brown waters towards the little patch of land that he owned. Slowly he dragged and lifted each log into place, often raising pieces of wood weighing many tons using a series of pulleys and winches.

I finally pulled my rain-filled canoe up onto the swampy beach next to a small clearing and Colbert's wild home. To my total surprise, it was a breathtakingly beautiful place. The water-blackened wood contorted and twisted into fantastical shapes, the cabin looked like a surrealist's interpretation of what a cabin could look like. She wasn't as pretty as some cabins I have visited over the years, but she exuded life and happiness and love.

Colbert's dedication to the task was half the charm. How many of us build our own houses? Very few, I'd guess, and of those who do, how many have actually built it with our own hands. Constructing a home with blood, sweat and tears. Most of us live in a home, a flat or an apartment built by someone else, but cabins are homemade. I suppose like the difference between an apple pie made at home and one bought in the shops, part of the beauty of the taste is in the making. Thought and love and time have been put into that apple pie.

Time. Isn't that what it's all about? Of the 60 wild men and women I have been fortunate enough to meet over the years, not one of them has worn a watch, risen to an alarm clock or been a slave to time.

How often is time a distraction or an excuse? If you think about it, almost all of life's premium products have attained extra value because of time. The handpicked grapes and the years of maturity of champagne. The handmade craftsmanship of haute couture. The premium paid for a jumper, lovingly hand-knitted over eight months.

Time has a value. Time is money, as they say, but time is also a luxury. How many people cling to the anticipation of the upcoming weekend? How many of us can't wait for holiday times or the bank holiday weekend, when we have … time? As a reaction to the premium value placed on that 'time', we have created a culture of cutting corners and of trying to be more efficient, often to the detriment of our physical and mental well-being. When did time become such a premium? Do we live to work or work to live?

The simplicity of cabins sometimes extends to their inhabitants. Our earliest pre-conceived ideas for the TV series – that the people would be 'simpletons', or bearded, washed-up hillbillies castaway into cabins with their pot and their guns – while not completely wrong were still way off the mark. By and large I have found the opposite to be true. Most wild folk have more time to be considered. They have time to think, absorb and reflect.

So often the relationship between man and the wild is one of exploitation. We destroy and plunder, but spending time with Colbert and John had shown me that we can be more sympathetic. It would be remiss not to mention the many people and tribes around the world who still live in harmony with their surroundings, but John and Colbert had demonstrated to me that we all have the ability to rethink our relationship with our environment.

One of the earliest and most memorable of my 'wild men' encounters was with Dave Glasheen, a one-time multi-millionaire stockbroker, until he lost it all, which was the catalyst for a massive lifestyle change. He swapped his Sydney mansion for a remote, uninhabited island. For more than 20 years, Dave carved out a life alone except for his beloved dog, on Restoration Island, or 'Resto' as he liked to call it.

Dave was the quintessential castaway. With his long white beard, mad hair and tattered shorts, he looked more like Robinson Crusoe than Crusoe himself. He had also become a celebrity, his face splashed across newspapers, not because of his castaway status but because he was taking on the government in their attempt to evict him from the island. Dave wasn't prepared to go without a fight and the spectacle of this bearded castaway standing up to the state, with its echoes of David and Goliath, created a minor media frenzy.

He never stopped talking. From the moment I arrived on his island, until the moment I left a week later, he barely paused for breath. I wondered whether it was the result of isolation and a pent-up craving for company. The female mannequin that sat in his living room seemed to confirm this, like Tom Hanks's volleyball in the Hollywood film version of *Cast Away*, only weirder.

Dave was certainly eccentric. He often walked around the island naked (who wouldn't if they lived alone?) and he had plenty of conspiracy theories. Although he was isolated from the mainland by crocodile- and shark-infested waters, he remained largely connected to news via an old satellite dish that allowed him to listen to the radio.

What surprised me most was his happiness and contentment. While any mention of the court case was enough to turn him puce, for the rest of the time he was always laughing. We drank homemade beer and went fishing. I helped him with some minor repairs and we met some of the local Aboriginal community with whom he was working in his efforts to remain on the island. The Aboriginals claimed the island as indigenous land and they too were fighting the nation for the rights of the island. Dave had joined their fight in the hope (and promise) that once transferred to the native Australians, he would be given permission to remain as a guardian of the island.

The island itself was relatively unremarkable as far as islands go, although it has a pretty remarkable place in history. It was here, after being cast adrift in a lifeboat by mutineers, that Captain Bligh of HMS *Bounty* and his officers finally 'restored' their meagre supplies. Obsessed with the story of the *Bounty*, I was star-struck to walk the island and imagine I was treading the same sands as Bligh, who had endured so much.

Palm trees and sand and, well, that was about all. As I say, it was unremarkable, but for Dave it was his home.

Dave left a great impression. His lifestyle change had been one of the most dramatic: from millionaire to castaway. What struck me was that he didn't miss his former life. He had adapted effortlessly to island life and had become one with the landscape. He had a synergy and a symbiotic relationship with his surroundings. He only ever took the fish he needed from the ocean. He was deeply respectful of his environment and the island wilderness.

Once again, despite the comparative difficulties of solitude and isolation, this lone castaway had found more happiness than he had ever experienced with money. Does money ever buy you happiness? It can help, but it is certainly not the magic bullet. So many people spend their lives dreaming, hoping and working for money in the pursuit of happiness, but at what cost?

I suppose it depends how much money we are talking about. There are plenty of people who sadly don't even have enough for survival; those souls born into poverty or thrown into it by either a tragic twist of fate or an unfair system. I have long held a guilt over my own privilege. Every time I spend time with those less fortunate than me, I am reminded of how lucky I am and how much suffering there is in the world. But I have also seen the other end of the spectrum: people who are born into or inherit or win great fortunes and whose lives spiral out of control. Too little money is terrible. Too much money also has the power to destroy. You'd be amazed how many unhappy billionaires there are on this planet. I don't feel sorry for them, but it's important to realise that money will not cure all. It is not the key to happiness.

Dave had known both money, a great fortune of many millions, and he had also known nothing, and he had been far more content living a subsistence life than that of a Sydney millionaire. The time I spent with him promted me for the first time to really question my relationship with money and the commercial world. Most of us are slaves to money. We have dedicated our lives to the making of money so that we can buy stuff. Things we don't really need. I would go on to spend time with Mark Boyle, also known as the moneyless man, and also an American couple who were also once millionaires but gave their fortune away and moved to an adobe mud house in Morocco. Each of them made me question my own monetary principles.

Another memorable wild man was Dan, who lived in a small hobbit house in the prairie meadows of Oregon. Dan truly lived the simple life and had restricted himself to the bare minimum of items. He lived in a little hole dug into the ground. To get in required crawling through the two-foot-high door on your belly. Inside was a tiny gas stove. One pot. One pan. One plate. One knife. You get the picture.

Of all the wild folk I met, Dan seemed to be living the closest to wild perfection. His impact on the landscape around him was negligible and he had found simple happiness. He was another of those solid, good karma people. There was such calmness amid the simplicity of his life. We sat in his small sweat lodge, heated by the wood from the fallen trees in the meadow, until I couldn't take the heat any longer. It was like the toxins of accumulation were being purged from my body. I leapt into the chilly waters of the nearby stream. We caught a small trout and ate it for our dinner.

We were in a wild place but surrounded by humanity. All around us, Oregon life continued, with its SUVs, Big Macs and giant milkshakes. All around us, people had committed themselves to the burden of debt in return for a house and a car bigger than they needed. Tethered for life by the stranglehold of debt.

Every time I come back from an expedition, a journey or a life in the wild, I am determined to simplify my life and sever my ties with the 'system'. That would be the system that makes us believe we should all aspire to go to university and leave with £50,000 of debt. The same system that encourages us to get a job to repay the debt. Then we are told we must 'own' a house or flat, which requires more debt. A mortgage. We are sold loans to buy the latest fashion and technology and car and holiday. The debt mounts and, before we know it, we are slaves to the banks. The economy has enslaved us, and for many people that burden of debt can be impossible to escape. We live our lives pursuing money. And then we die.

It sounds a little brutal when you condense it like that, but I'd hazard a guess that many of you reading this book will recognise the 'goal' we set ourselves. You see, we all buy into it. Myself included. We burden ourselves with financial debt to lead the life we think we need. Of course, it's not as simple as that. Throw into the picture jealousy, showing off, one-upmanship, material obsolescence and the whole commercial game, and we have a much more complicated picture that effectively renders most of us slaves to money.

I have tried and always failed to escape this vicious cycle. Every time I commit to the simplification of my own life, I am seduced by the power of commercialism. The phone with the better camera, the new pair of walking boots, the outdoor jacket in that orange that I had always wanted. You see, I'm one of you still, not one of them. I'm desperate to join Dan's club, but the answer is that I don't think I have the strength of character.

I also have commitments. I have already introduced my children, Ludo and Iona, to the 'system', and to withdraw them now would be unfair. For once you have tasted commercialism, it is difficult to forego it.

There are always a million reasons why we shouldn't do something and maybe only one reason to do it. Perhaps that's why it often takes something unexpected and life-changing for people to take the leap of faith. Once they have broken the seal of doubt or fear, it is easier to dive in.

I have never had the confidence. Too many what-ifs. And most importantly, family to consider; not just my immediate family, but my extended family. Both Marina and I are very close to our families. We all live near to one another and we see each other daily. I suppose we have quite old-fashioned family values and I often wonder whether I could ever abandon family for lifestyle.

The recent coronavirus lockdown showed how fragile our system is. So many people and businesses almost live on a hand-to-mouth basis. What we witnessed was an economic collapse after just a few short weeks. That would never happen in the wilderness. Even if you were tent- or cabin-bound for a couple of months, chances are you would be able to forage enough wood, food and water to survive. Of course that is a very simplistic analysis and there are plenty of variables, but my point is that we are far freer once we have severed our ties to the pound or the dollar.

But it doesn't have to be this way. I have seen the light. I am aware of the hypocrisy of this coming from someone who has been able to make enough money to buy these things without the borrowing and the debt (that isn't entirely true by the way; I have a massive mortgage), but I have met dozens of people who have proved that it is possible. And if you take yourself out of the environment, you break free from the 'system' and find real freedom. You'd be amazed how many of us are slaves to our lifestyle. Fear of losing what we think we need, or what we think makes us happy, is all that is stopping us from slowing down.

This period of coronavirus isolation has also given people a little glimpse of a simpler life, and a lot of people have tasted 'freedom'. Just like some of my wild folk.

Dan had surrendered to simplicity and with it had come untold happiness, contentment and wisdom. I spent hours listening to his tales and thoughts on the world. No finger-wagging, no accusation or blame, just gentle observations on life and how to be happier and freer. Dan demonstrated that if you unshackle yourself from the expectation of society, you open up the possibility of being so much happier.

* * *

Perhaps this 'unshackling' is the reason why many of the wild folk are ostracised or estranged from their families. The majority of those who I met were alone, but I did meet some groups committed to living wild. Of these, it was the Long family who would make the most profound impression on me. Often described as New Zealand's most remote family, the Longs live in a cabin made of driftwood, on a lonely beach a three-day walk from the nearest settlement in Milford Sound on the South Island of New Zealand. To say they are remote doesn't really do it justice, and it is this remote isolation that gives them their uniqueness.

For three days, my camera crew and I trudged through wild ferns and grass, waded across glacial rivers and hopped across boulder-strewn beaches, until finally we reached a hidden gorge, on the other side of which was a tiny wooden cabin.

Robert Long and his Australian wife Catherine had moved to this wild, lonely place several decades ago, and it was here that they had raised their two children. The children have now left the nest but their upbringing was in its truest sense in the wild. They fished fresh crab and lobster from the wild, untamed ocean and they killed possums, a pest in New Zealand, for their meat and their fur.

Life was as simple as you could imagine. They grew their own vegetables and the children were home-educated. To offset their living costs, Robert would carve the jade he found on the local beaches to sell. An artist, he would also paint pictures that he would sell in galleries around the world. His art commanded quite a price, partly I'm sure because of its provenance – the fact that he created these pictures from this tiny little cabin in the middle of nowhere.

A kind, warm and engaging soul, Robert often welled up with tears when he talked about his children, even while they were still there. I envied their family companionship and the closeness of their unique dynamic. When I talk of wilderness living, I often refer to the Longs as the essence of that spirit. The location. The cabin. The disconnection. The nature. The landscape. The minimal footprint. The love. The life force. Their life was exactly the sum of its parts and to an outsider like me it looked as close to perfection as I have ever known. I could nit-pick and complain about the sand flies or the prolific rain, but then that would be undervaluing the perfection of everything else.

The location of their little shack was idyllic; from their kitchen you could watch the Southern Ocean as it crashed onto the foreshore. If you were to make a Hollywood film about the perfect 'wild family', it would be the Longs. Kind, generous, interesting and interested, they had an effortless life in the harshest of environments. I spent a memorable ten days with them, living in my tent on their front lawn to give them a little space.

I would return to the Longs eight years later, in 2019. It was a fleeting visit. And I really mean fleeting. Time-squeezed by bad storms and diverted flights, I dropped in by helicopter. I had four hours to catch up on nearly eight years. Robert cried when he saw me. Those tears were infectious, 'I knew you'd be back,' he smiled, as he hugged me. I felt such a sense of belonging and happiness to be there again. Everything was familiar. Robert felt like family, like an extension of my father.

What I realised then was that all of my wild folk encounters have been as much about the people as the landscapes. I probably knew that all along, but while I often wax lyrical about the beautiful landscapes, those wild experiences were forged by each irreducible interaction of man and nature. Each person, couple and family had created their own unique synergy with the landscape. Sometimes it was beautiful, at other times maybe not so.

I have often wondered whether we become a product of our environment. If you meet a New Yorker, a proper Manhattan New Yorker, they have this frenetic energy that comes from the 24/7 pace of life from the city that never sleeps. If you meet a surfer from southern California, they are often more laid-back. Of course I am stereotyping people and cultures, but by and large the landscape in which we all live becomes a part of our character. Adaptable and chameleon-like, we change to suit our environment. We become a part of that landscape.

But when we move ourselves between environments, what happens to those emotions? When we go on holiday, do we leave our cares behind or do we simple bury them deeper? Are emotions a bag we are always carrying, or are they a wrapping that we can discard? Well, I'm no expert on the subject, but like everyone I have highs and lows. I have good days and bad days.

Some of us are adept at hiding or disguising our emotions while others wear their heart on their sleeve. I think it is irrefutable that we carry our feelings and emotions wherever we go. But what about happiness and complacency? Another question my encounters provoke is: when my wild folk head into their uncomplicated lives, do they take their unhappiness with them? Is it a symptom of their lives or the world around us more generally?

There are plenty of examples of people who have suffered a crisis in their lives and have run away from it. They have gone to a faraway place to escape it or to try and absolve themselves from the responsibility. The problem is that we can never really escape from ourselves, and from that loud voice within. Today, we are suffering a worldwide mental health epidemic. Clinically diagnosed depression has never been higher. Young men and women are taking their own lives. That doesn't sound to me like escaping from our psychological crises.

If, as is my hunch, the wilderness has the power to heal our mental well-being, why aren't more people doing it? We've all had that revelation when we go on holiday. After a few days we feel the weight and burden lifting as work disappears from the forefront of our thoughts, but it never goes away. If you split from a loved one, you might avoid heartache for a couple of weeks or even a couple of months by taking off to a faraway place, but eventually you're going to return to the reality of life. You can't always run away from life, but you can run away from the ills of society for a short while.

Almost all of the wild folk I have encountered over the years were compelled to change their lives after some seismic event had shaken their world. Bereavement, loss, illness, financial ruin, but occasionally it was all part of an experiment. Although life is often happier in the wilderness, it isn't necessarily easier or more kind; on the contrary, it is often harder and, well, wilder, but then that is the unofficial deal: the unwritten contract between wild folk and the wilderness they inhabit.

I think it's the fact that you have to work harder that makes it so appealing. It's part of the draw. The satisfaction comes from the gap between effort and reward. In cities, the distance between effort and reward can sometimes be negligible, but in the wild, it can be vast. The greater the gap, the bigger the reward, or perhaps

I should say, the bigger the sense of reward. You see, the sense of satisfaction comes through the effort. It's the reason I felt so elated after rowing the Atlantic. I could have rowed across the Channel, an effort in itself, but the feeling of satisfaction would be fractional.

I am often fascinated by the choice of interactions taken by the wild folk with the wider community. Some people are content with their own company while others require constant companionship. I enjoy other people's company but I am also happy alone.

How much stuff do you own? Really, think about it. I don't know about you, but if I include every knick-knack, knife and spoon, and everything I've accumulated over the years, it is probably in the tens if not hundreds of thousands of items. I am a magpie. I gather what I consider to be beautiful or sentimental things. I like to think I could remember all of them, but there is no way I would remember that tiny souvenir from Kenya, hidden in a long-forgotten drawer.

I think it is this hoarding and gathering that is, in part, causing us so much angst and the world so much damage, because we really don't need much. Society makes you think you do by bringing out seasonal launches of clothing or technology with in-built obsolescence, but we don't really need it. There is a movement to reduce your worldly belongings to just 100 items. I've thought about it, fantasised about it, but I've never been brave enough to do it. My son's Lego collection would break the rationing alone!

* * *

One of the common traits among the wild folk is that many are loners. Like Nikola Boric, who embraced the wilderness as a challenge to test himself. Nikola was a former Croatian athlete who had dropped out of the system. He had grown up during the wretched, terrible, bloody, pointless war between the Serbs and the Croats. I'm not going to give you a history lesson now, but it turned neighbours against one another.

While I was at university during the early nineties, I joined an aid convoy that we drove overland from Portsmouth to hospitals, schools and orphanages across Croatia. It was my first experience of the modern battlefield. We drove through towns destroyed by shelling, past the charred remains of buildings burned because of their owners' ethnicity. I was horrified to see how quickly neighbours can turn on one another once the people stir the hornet's nest of identity and belonging. This tribalisation of ethnicity is a slippery slope and the result was genocide.

I hadn't been back to Croatia for nearly 25 years, and I was horrified to see that in a quarter of a century, the burned-out, abandoned buildings remained. Graffiti with 'Serbs out', and 'Croats live here' was still visible on the sides of inhabited buildings. Shell, mortar and gunshot markings still pockmarked the landscape. If you didn't know it, you'd be forgiven for thinking the war had ended just a few years ago.

Among all of this, Nikola Boric had become a world-class athlete, competing in his own right in cycling, swimming and eventually triathlon. He had become a medal-winning international champion, who eventually swapped the track for the field of management and training. He coached footballers and triathletes and he took them to world championships, international tournaments and the Olympic games. He became a celebrity in Croatia, but happiness eluded him and one day he dropped off the grid. He packed a rucksack and hiked into the woods north-east of Zagreb where he hung a hammock and spent the winter. Let me tell you, a Croatian winter is a hardship. Minus 20 degrees C temperatures and plenty of snow tested Nikola's resilience and backbone, but then that was the purpose.

Nikola had been used to suffering for his sport. He was used to pushing himself in the pursuit of excellence, and now, here in the cold winter of the Croatian wilderness, he was testing his own spirit and resilience. I suppose it was a kind of flagellation. He needed to suffer to see what he could withstand. He was looking for direction and a calling. He had become despondent with the professional sporting life, which he felt had sold out to materialism. It had become a world more impressed with money and sponsorship than with sporting excellence.

The wilderness offered an escape where Nikola could rebuild his disillusioned life. Wild folk are commonplace in many corners of the world, but Croatia had never encountered such an eccentric citizen who had abandoned his showbiz life for a hammock in the forest. He became a celebrity for his eccentricity. The locals in the woods where he was living were unimpressed; a religious community, they thought that he had succumbed to paganism. Nikola found copies of the Bible hidden among his meagre belongings, placed there by worried locals to help him find the correct path. After a winter in the hammock, Nikola decided it was time to have a home. He dug away at the earth, and built himself a house of straw and mud. He shifted tons of earth, mud and sand to create an adobe cabin in the woods. A refuge from society.

With his big, bushy, black beard, his sportsman's physique and his bald head, he looked imposing. The kind of person you might avoid in a dark alley. I'd guess that his appearance was part of the suspicion he aroused in others.

After a while, Nikola expanded his hut to create a smallholding. With true sportsmanlike drive, he didn't just grow any old vegetables and crops, he made sure he grew the finest vegetables that Croatia could grow and sold his produce to the finest restaurants in the country. In return he got the small amount of money he needed to survive.

And of course, the former swimmer also dug himself a 100-metre pool. It wasn't the finest pool I had ever seen, and when I visited it had been overrun by the pigs, but it was an impressive sight and you can't help but admire the hard graft and hours that went into the digging. It was the blood, mud, sweat and tears that drove Nikola. He wouldn't settle for anything less. He could have made his life a whole lot easier but he didn't want to. He enjoyed the thrill of the challenge. This brings us back to the effort over reward graph once again. Nikola knew from his sporting days that the bigger the effort, the greater the reward, and the result was incredible.

I often speak of spirit when I talk of people, and wild folk in particular. Once you have spent time with extraordinary people you start to pick up the hint of their spirit and the motivation that drives them. Nikola had been angry and bitter. I could sense it was still there. Maybe it was the reason for the toil – to purge the anger from his body.

It surprised me how intolerant the local community were to this slightly odd man in the woods. When he met his girlfriend,

Zorana, she had been warned to stay away from him, by her own family. I had seen this resistance to people like Nikola before, but was surprised he caused such irritation by living such an inoffensive life. It wasn't as if he was rubbing people's noses in it – he was living far away from other people. But what I have discovered over the years is that people are afraid of the unknown. They are suspicious of others who break the mould and follow their own path.

The wilderness doesn't judge. The wilderness in all of her forms – the jungle, the mountains, the desert, the open seas, the ice cap – brings out the real person within. It allows us to be the person we want to be, rather than the person others want us to be.

Just as in urban life, wilderness living has plenty of flaws and imperfections, but at the heart of the narrative is the power of the wilderness to bring out the best in all of us. Where modern society has a habit of dividing and bringing out the ugly side in our increasingly polarised world, the wild, natural wilderness remains true to her core values. Respect her and she'll respect you. I have seen it many times, and with that freedom of expression comes the liberty and happiness that I've seen in every individual I have had the privilege to meet over the years.

For Emma Orbach, Brithdir Mawr, her small eco-community hidden in the Pembrokeshire countryside of Wales, is a spiritual place where nature is not just respected but revered and worshipped. Emma, an Oxford graduate, set up the community with her husband, architectural historian Julian. She has spent 13 years living without electricity in her self-built roundhouse made from wood, mud and turf. It looks more like a hobbit house than the traditional Welsh architecture.

For the first five years they enjoyed the simple life, until an aerial survey spotted their lost tribe and they were plunged into a decade-long battle with officialdom, the bane of all wild folk. This 'lost tribe' in the Welsh countryside was home to an unknown number of free-thinking, spirited folk in search of happiness in the wilderness.

Emma is a quiet, reserved, thoughtful individual with an intuitive connection to the wilderness around. We, the film crew and I, were asked to stick to the basic trails around the forest to avoid stepping on the flora and fauna; if we did, we were asked to apologise for our transgression. Among the roundhouses was one large place of worship, a mighty wooden structure, in which the community could hold ceremonies dedicated to the wilderness. Here the reverence for nature reached spiritual and religious levels. Neither pagan nor a cult, this small, shy community kept itself to itself.

During the few days that I stayed with Emma, I saw only fleeting glimpses of other folk living discreetly in the woods. Frightened away by the presence of our film cameras, I rarely saw another person, although I knew that hidden among the trees and the tall grass were dozens of other people. In fact, I would have sworn I was in a pristine, uninhabited oasis, were it not for an old-fashioned ring-dial telephone sitting on a chair in the corner of a field.

It was an extraordinary place. I liked Emma a lot. It was the first time I had met a community who revered the wilderness like a religion. They worshipped their surroundings as if it were a god.

For this small Welsh community, the wilderness had become a cult-like obsession. Emma had tried to create a community in harmony with the wilderness. Not only did they revere it and celebrate it, but they had become a part of the landscape.

I have become fascinated by the communities around the world who have built a spirituality around the wilderness – part paganism, part shamanism – which makes sense of a spiritual connection between them and the natural world. Have you ever taken time out to sit with nature? Most of us race through the wilderness. We incorporate it into our busy lives and we limit the time we spend there, but if you just slow down, then you really begin to appreciate what is around you. All of the wild folk I have visited have had the luxury of time to let nature absorb them into her bosom. They are much more finely tuned to the nuances of nature. How often have you walked through a seemingly empty wood or forest? Our heavy footfall scares anything natural away. But when you settle and sit still, soon you begin to hear and see things that would otherwise be invisible. It's as if the wild encourages us to be calmer.

I suspect this is one of the reasons I have experienced so much happiness among the wild folk who have traded in the unnatural world for a natural one. I have always been attracted to spirited folk, the non-conformists, the optimists, the fearless pioneers, the shepherds rather than the sheep, and I've been fortunate to meet hundreds of fascinating people throughout my life, but there is an extra allure when people dedicate themselves to looking out for others.

While it is easy to focus on what we are doing wrong to the planet, we often overlook the selfless, life-changing work being undertaken by spirited souls all around the world. In many ways, we are all products of the people we encounter and meet. We are the sum of so many things, from the experiences we have to our environments.

As for the wilderness, here too we are forced to adapt to our circumstances, the difference being that if we fail to conform, we really might die. Unless you protect yourself from the cold, or the heat, or the flora and the fauna, then chances are you won't make it. Where the cities force us to adapt in a competitive manner, in the wilds, we are forced to change for our very survival.

A few years ago I teamed up with the Swedish tourist board to do a wilderness experiment in which we took a selection of urban folk and tried to clinically prove the medicinal healing qualities of the wilderness; in their case, that of the Swedish forests.

On a tiny island in the south of the country, we built a series of glass cabins. We were inside, but outside. They protected us from the weather, but allowed us to be a part of the fabric of the forest. Among our guinea pigs were a Paris taxi driver and a New York events organiser, both pretty stressful jobs in fast-paced cities. Before the experiment began, each of our volunteers underwent a series of clinical tests that included the obvious heart rate and blood pressure but also psychometric testing that included speed of response, reasoning and reflexes. With all of this information we were able to create a baseline from which we could track anxiety, mood and stress.

We ate breakfast around the camp fire. We paddled around the little island in the lake. We did plenty of walking and we used the wood-burning sauna before leaping into the chilly waters. We slept. Sometimes we did nothing. A little like those cabins themselves with their glass walls, we soon began to be absorbed into the forest all around us. Silence and solitude soon took over from the speed and haste of city life and we all, myself included, became less stressed.

The data was astonishing. The wilderness appeared to heal our city anxieties. Resting heart rates reduced, as did blood pressure. Through a series of tests, researchers established that our reactions were faster and our memory had improved.

So here we had proof that the wilderness has the power to heal our physical and mental health, but what about the effect on us of those we meet? It's not a great leap from here to see that our chameleon-like tendencies to adapt extend to the people that surround us.

Take the 'tribes' of cities. The cool hipsters, the bohemian hippies, the city slickers. Each part of the city breeds a certain type of person. Perhaps they have chosen that region or area because it is already a part of their character, and we do find ourselves drawn to like-minded folk, but there is also an undeniable tendency to change according to our surroundings. And herein lies one of the problems of modern society: we have created vast, anonymous, online communities that foster more than just hipster beards; they can bring out the worst in people as we scramble for a sense of belonging. The result is the ever-increasing dominance on social media of intolerance and discontent. Hate and fear, loathing and outrage, extremism and wokeism are rife.

Imagine, then, if we only ever surrounded ourselves with incredible people. Imagine if we could absorb the karma and character of those spirited people we encounter. I have met so many amazing people throughout my journeys around the world and every one has had an impact on my life in so many ways.

The relationship between man and the wild is complicated. En masse it is hard for us not to exploit and abuse it, but individually and in small groups there can be a beautiful synergy and collaboration. Living and spending time in the wilderness creates a respect. The wild folk will experience the 'cause and effect' of their actions first hand, something most city folk are entirely oblivious of. Wild folk will identify with the smallest of changes in their environment. They become highly sensitive to changes in nature and highly perceptive to the effects of their actions.

The wilderness can provide a generous hand-to-mouth existence and in return the wild folk become the guardians and the protectors. There is a notion that to live in the wilderness is a battle between man and wild, but the reality is that it is a beautiful serenade in which both can benefit from one another.

The decade spent with wild folk around the world reminded me that the wilderness is much more than just somewhere to explore and endure. With the respect and care of its inhabitants, it can be a place of great happiness, a place of hope and rehabilitation. And a place where the simpler life that some of us crave can be found. As Dan in Oregon once told me, 'If you want nothing, you have everything.'

CHAPTER EIGHT

WILDLIFE

'Let Nature be your teacher.'
William Wordsworth

The wilderness would be nothing without its wildlife. It's the wild animals and plants that bring the life, the colour and the drama to the wilderness. But who are the true custodians of the wilderness? The flora, the fauna, or humans? We all know that it should be shared between us fairly, but reality rarely works that way. The hierarchy of the world is an extraordinary one. I don't need to tell you that human beings are the most dangerous animals on earth. As the alpha predator, we often live with an arrogance of certainty that fosters selfishness. It was Richard Dawkins who wrote in his book, *The Selfish Gene*, about this very trait.

Whenever that certainty is questioned, we recoil in fear and often retaliate in blood. Each year a handful of people are killed sharks, polar bears and lions, but whenever tragedy strikes, those deaths are reported excitedly on the front pages of newspapers around the world. When a human is taken by an animal it is an affront to our superiority and, more significantly, a reminder of our vulnerabilities.

Man's dominance can often have negative repercussions, but what about the natural inhabitants of the wilderness and what can we learn about their relationship with the environment? Animals don't hate and we are supposed to be better than them. Just take a moment to think about that. Superiority is all in the eye of the beholder.

Ever since I was a small boy, I have loved animals. The son of a vet, it is probably unsurprising that I have been around animals all my life. From the childhood pets that included parrots, dogs, cats and guinea pigs to the wild animals I have been fortunate to work with over the years. I have tracked blue whales off the coast of Sri Lanka, and I have filmed great white sharks hunting in the wild waters of the Farallon Islands in the Pacific. I have followed the migration of the wildebeest from Tanzania to Kenya, and I have been lucky enough to live with chimpanzees in Uganda.

Each encounter has taught me about our relationship with animals and helped me to interpret my own base human behaviour. Each animal has shown me the importance of understanding that it is wildlife *with* man, not wildlife *versus* man.

You never forget your first wild elephant. The sunrise had painted the landscape a golden hue as our Land Rover bumped its way along the track, leaving a mighty cloud of dust in our wake. Huge acacia trees stood proud across the plains, like ships' masts casting long shadows from the rising sun.

An African sunrise is unlike any other. It feels like a theatrical performance as the thin layer of dust in the air changes from yellow to orange to ochre to gold. Goose bumps covered my arms and legs from the chill night air and I willed on the warmth of the sun as it lit up the savannah before us. We passed through an area of thick vegetation and slowed down, and there before me was a herd of about 50 elephants grazing in the tall grass. I had seen lonely elephants at London Zoo, but had never seen a real elephant in its natural habitat, in the wild. It quite literally took my breath away. The sun cast its warm glow over our vehicle, but the goose bumps remained, not because of the cold but because of the 'moment'.

I wanted to laugh and I wanted to cry. The majesty of these incredible beasts, so unlike anything else I had ever seen. The smell, the colour, the silence, the mystery, the wonder. I don't think I had ever experienced such a cocktail of sentiments. Growing up in Central London, I was lucky to see a pigeon or perhaps an urban fox, but here I was, watching a mighty herd of elephants. I was amazed at how little noise they made as they ate and walked through the tall grass, baby elephants following close behind.

I was 25 and I was in Zambia for *Hello* magazine, which had asked me to do an 'At Home' feature. Never comfortable inside, I suggested a series of 'In the Wild' features, which took me around the world from East Timor to the Arctic Circle and now, here, to Zambia and my first experience of Africa.

I don't want to sound too clichéd, but I have never known a morning in Africa when I haven't woken up happy. It really is a joyful place. Once Africa has got under your skin it is hard to forget her.

* * *

I have been fortunate to have had many great elephant encounters since then, but I will never forget the most recent one. I was in Kenya filming a series called *The Great Migration* in which I spent a year following the most spectacular journey on earth, as more than a million wildebeest headed up through the Tanzanian Serengeti and into the Kenyan Maasai Mara. We were in southern Kenya, close to the Tanzanian border, and close to human settlements and agriculture. We had set out early in the morning before sunrise to catch the migration on the vast open plains. As we were heading through a dense copse of trees we came across a mighty elephant. We could see the unmistakable grey skin hidden behind the vegetation, but she wasn't moving. Something was wrong.

Carefully we got out of our vehicle and we approached. There she lay, in a pool of congealed blood, her tusks hacked from her face. Thousands of flies buzzed in a feeding frenzy. Apart from her face, she was intact. Perfectly preserved. From behind she could have been sleeping, except that she wasn't. She had been killed for her ivory.

I am not an angry person. Some people are born angrier than others, but for me it is a rare emotion usually inflamed by unfair, wrongful, or discriminatory behaviour. There is something about mean-spirited spitefulness that makes me see red. I sat there on the ground, my hands on my face in despair. Tears soon turned to rage. Bitter, seething rage unlike anything I had ever felt before. How could someone be so cruel and selfish to kill a beautiful, majestic elephant for its tusks?

At school, I only ever remember feeling angry when I was accused of something I didn't do. I'm not sure why it elicited anger rather than sadness but it always made me cross. It seemed so unfair. Injustice and inequality still make me cross. I have seen plenty of it over the years. Humans can be quite cruel when it comes to their fellow man, our relationship with animals, and the planet.

It was Richard Dawkins who wrote in *The Selfish Gene* that by our very nature we are quite a selfish species, especially when it comes to our relationship with the wilderness and nature. Humankind has raped and pillaged the natural world for her riches, and sometimes she bites back. I am a conservationist and an environmentalist and have been so ever since my year on Taransay, but I am also part of the problem. I can see the irony and the hypocrisy in my spirited fight when I myself have a negative impact with my own carbon footprint.

But the conversation about the environment has changed and in recent years it has, like so much else in life, became increasingly polarised. There is a view that you can no longer stand up for the environment unless you are GREEN! – not just green. There's a difference. Doing your best to be green no longer cuts it. Raise your head above the parapet and you'll be accused of gross hypocrisy unless your green credentials are impeccable, and as a traveller and a journeyman, mine are clearly not.

The planet is in crisis. We are at breaking point and the world is divided into those who deny we have had any impact and those who dedicate their lives to challenging the old ways. The rise of Greta Thunberg and the birth of Extinction Rebellion came about with the election of climate-change-denying Donald Trump.

People are confused. What is really happening? Who do we believe? What can we do to make a difference? This is meant to be an inspiring book full of positivity and hope, but it would be remiss of me if I didn't touch on our relationship with the natural world. I have shared with you a number of lessons to take from the wilderness, but what about our physical relationship? What about the impact we have *on* the wilderness: the flora and the fauna?

I have rarely met a selfish community or individual who is immersed within the wilderness or natural world. First-hand experience of cause and effect is one of the most important and powerful ways to appreciate our impact. Once again, I should say that I can see the hypocrisy in my own situation. I fly a lot. Too much. My carbon footprint is higher than most in the air miles required to reach some of the wildest and most remote places on earth. There is a dreadful irony in that, I know. I have thought long and hard about this for many years now. Of course I carbon offset my travel, but that is really just a way of trying to justify my carbon footprint to myself, and it isn't really focusing on the problem.

I have always considered myself a positive campaigner. I have never felt it effective to 'guilt' people into change. I'd rather empower them to go and make the change happen, and that is what I hope to do here.

You can never overestimate the power of nature. Over the years I have had plenty of first-hand experiences of that power. I witnessed the devastation of Haiti, after the earthquake that rocked the island in 2010 killed an estimated quarter of a million people. Visiting the island a couple of months after the quake, I was shocked and horrified on witnessing a population still struggling to pick up the pieces. Three and a half million people had been affected by the quake and the country's infrastructure had been destroyed. It was one of the few times I have visited a place in which I saw no hope. I hate to admit that, but I genuinely couldn't see how this tiny nation could possibly pick up the broken, shattered pieces and rebuild. Almost every building including the presidential palace had been toppled. Debris and litter were spilled into one seemingly never-ending pile of wretchedness. While I was there, a cholera epidemic spread across the island, claiming even more lives and bringing further despair. Among the devastation, teams of international humanitarians were trying to do the impossible and help the battered nation get back onto its feet.

On the other side of the world, I visited Japan, a year after the terrible tsunami claimed upwards of 20,000 lives in March 2011. Generated by a powerful undersea earthquake, the tsunami all but obliterated the residential area of Natori, in the Miyagi prefecture in northeastern Japan. By the time I visited at the end of the year to make a film about the power of water, the impact was heart-breaking. I had never seen such devastation, with boats in trees and ships marooned on highways many miles from the sea.

But already the clean-up had begun: there were neat piles of buses, cars, bicycles and motorbikes. Each one had been labelled and documented. Where vehicles had been broken apart, their remains had been pieced back together like jigsaw puzzles. It is no surprise that this was Japan, the same nation that invented the beautiful art of kintsugi, which reassembles broken porcelain as an art form.

There was a deep and heartfelt respect for the lives lost but also for Mother Nature. Japan has long held a reverence for the wilderness. It was also the Japanese who invented the art of forest bathing, known as shinrin-yoku or 'forest medicine'. It literally translates as 'taking in the forest atmosphere', and it has become a cornerstone of preventative health care and healing in Japanese medicine. The idea is simple: the forest has a soothing, calming, rejuvenating and restorative effect on the body. In Japan, shinrin-yoku has been clinically proven to boost the immune system, to reduce blood pressure, to improve the mood and to increase energy and improve sleep. Indeed there are many cultures that use the natural world for her healing abilities, but the Japanese have a particularly humble reverence, which made the devastation of the tsunami all the more shocking.

In recent years, I have become a patron of the British Red Cross, and in that role I have visited a number of other places affected by natural disasters around the world. Shortly before climbing Everest, Victoria Pendleton and I visited communities devastated by the Nepalese earthquake of 2015, which killed nearly 9,000 people and left another 3.5 million homeless. Then in 2019, while on a filming assignment in South Africa, I made a detour to Mozambique, ravaged by Cyclone Idai, which destroyed the city of Beira and left countless people missing and dead. It was extraordinary to be on the ground as the humanitarian recovery mission scrambled into place. Teams from Non-Governmental Organisations all around the world gathered with their resources at the shattered airport. Each NGO was tasked with a different role. The British were in charge of latrines, the Italians handled medical aid, and the Spanish were responsible for putting up the tents for the homeless.

Despite our differences we have a marvellous ability to come together when the going gets tough. It's why I believe in community and why I also still think there is hope. Of course natural disasters are nothing new, but their increasing ferocity and frequency should set alarm bells ringing. The future will only look bright if we all manage to pull together for the common good and respect our environment.

I have spent a great deal of time investigating our impact on the land. In 2019, I spent a number of months exploring South Africa, Uganda, Kenya and Namibia to try to understand the implications of our actions. Our relationship with the natural world is complicated at best. But let's begin with our equally complicated and controversial relationship with animals.

I used to eat meat. I don't hunt and I don't like hunting, but then neither do I like abattoirs. It is probably the reason I am now vegetarian. I love animals and I don't think we should eat them on the scale that we do.

Even our diets have become controversial. I am often accused of hypocrisy because I have spent time with those who hunt and shoot and trap, but I have always prided myself on having an open mind. I am accused of sounding too worthy and smug when I say I don't like hunting; after all, I am frequently told, it is a tool of conservation. Animal stocks must be managed.

Let me tell you a story. In 2015, shortly after the news of Cecil the Lion being shot by an American hunter went viral to worldwide disgust, I found myself in a small farming community in Uganda. I was spending time with a subsistence farming family who wanted to show me how elephants were ruining their livelihoods. I watched in horror as an elephant came through the smallholding, tearing down fences and destroying all of the crops on which this farmer relied. He was in tears. Everything he had was lost. To an elephant.

Now your sympathy here probably depends on your attitude to who 'belongs' in Uganda. The people feel they have a right to work the land, while in the West we tend to think of Africa as the land of wild animals. Many in the UK will sympathise with the elephant rather than the farmer. It's the reason that animal welfare charities have far more money than charities supporting human beings. I think it's because of our feelings of vulnerability and helplessness that we often project onto voiceless animals. I am culpable of this myself. We often put animals ahead of people.

But back to that devastated farmer and his lost crop. He now has nothing. Nothing to feed his family. The crop was lost. There was no compensation. He can't claim income support, or even seek free medical help, because this is not the UK. What would you do if your family was starving and you had no place to go? Shoot the elephant and feed them with the meat?

He didn't actually do that. But he could. He might. Many of his neighbours had done it before. When I met a group of former poachers in Kenya and I asked them how they felt when they killed an elephant, one of the poachers said, 'I cried with happiness.' He was not thinking about the elephant but the money it would bring in to feed his children.

Many African nations have implemented rigid and strict laws against poaching, and if they are caught, poachers are punished with long prison sentences. But across the road, in a fenced 'conservation area', rich American dentists are legally allowed to hunt, shoot and kill an elephant for money. How does that sound when you are a farmer turned poacher? Why is it okay to kill an elephant if you pay, but not if you are starving? Let me be clear, neither is right, but you see the complications.

Over the years, I have spent time with people and communities who hunt for their food. One of the most shocking experiences was when I lived with the Hadzabe tribe in Kenya. They are nomadic hunter-gatherers and I watched them for a year as I made that documentary about the migration of the wildebeest.

Early in the morning, before the sun was up, the hunters headed off clutching bows and arrows and racing through the Kenyan bush. I am reasonably fit but I found it difficult to keep up. They had scented their quarry and they were off.

I caught up with them in a thick, dense area of forest. A hail of poisoned arrows were shot high into the forest canopy. Gravity did her thing, and soon poisoned arrows were raining down all around. One missed me by inches. There was no health and safety here. Then I heard the unmistakable sound of a large heavy body crashing to the floor. It was a mighty baboon. It was followed by another and then a third. It was shocking.

Within minutes, a fire had been lit and the baboons stripped of their fur, which would be used as clothing. The naked bodies of the baboons were thrown onto the fire, and an hour later, the Hadzabe were feasting. It was distressing to witness but was a little window into the world of those who still hunt for their food.

I know of many meat-eaters who would be repulsed by these baboon eaters, but just because our farm animals are reared and dispatched behind closed doors doesn't necessarily make it better. I might be a vegetarian now but I still don't like either method of slaughter.

But back to Africa and our complex relationship with wildlife. I saw crop-raiding chimpanzees in Uganda and fish-stealing sea-lions in Namibia, and all across the world I have witnessed a growing battle between human beings and animals.

Crocodiles, grizzly bears, sharks, polar bears, lions and tigers all have the power to remind us of our place, and any example of the animal kingdom's raw physical supremacy over us is catapulted into world headlines as we recoil in horror at a shark attack or a tiger mauling.

We might have just a handful of shark attacks around the world each year, but each and every one will be reported breathlessly in our international press. By comparison, according to the peer-reviewed journal *Marine Policy*, we kill an estimated 100,000,000.

Let me repeat that in case you think I've mistyped the figure: *one hundred million* sharks are killed every year. Where are those headlines?

Our fascination with sharks has led to a booming shark sightseeing tourism industry in South Africa, where dozens of boats compete to take voyeuristic tourists out to sea, where they are able to see wild great white sharks in the water from the safety of a cage. With ever-increasing competition for the valuable market, the industry has looked to control the situation. Tour operators need to guarantee a shark encounter. And the only way to do this is to 'chum' the water, which means basically pouring buckets of smelly and bloody fish guts into the water, often close to areas where tourists are bathing. The result is the habituation of sharks.

The same happens with polar bears and grizzly bears, which become habituated to tourists who accidentally or intentionally leave food for the bears in order to get the best selfie. The inevitable result is hysteria and the culling of magnificent creatures. It's pretty screwed up when you think about it.

To keep bathers safe in Durban, shark nets have been erected. The idea is that the shark swims up and down, window-shopping but never buying or consuming; but the reality is that sharks can get tangled in these nets and drown. They are a way of culling the number of sharks while also bringing a slightly misleading feeling of safety to nearby bathers.

The winner is the dollar or the rand. Tourists continue to flock to the beaches and the boats, and sharks continue to die. Of course, the shark just needs to swim underneath or around the net, and they do – but whatever you do, don't tell the tourists.

At the heart of this is the inherent fear that we have of predators. Countless films have been made in which sharks and crocodiles are the bogeymen, stalking and hunting us humans. If we added up the total number of people killed by the world's top predators each year, I'd hazard a guess that more humans are killed by chairs or toasters or puppies.

But we like to be scared.

What if we could change the narrative and change the perception? Maybe we would begin to respect these incredible animals rather than vilify them, and highlight their beauty as well as their teeth, jaws and claws. Unfortunately, there are communities who are forced to live as neighbours with some of these predators. A little like the farmer who lost his entire crop to a grumpy elephant, there are plenty of communities who lose their sheep to a lion or cheetah or who lose limbs or even lives to crocodiles. All across Africa, and in many other parts of the world, people and communities are at war with wildlife. I have met farmers in Namibia who kill cheetahs to protect their goats, and fishermen on the Skeleton Coast who kill sea-lions to protect their fish. I have also met farmers in Kenya who poison lions to protect their sheep, and villagers in India who kill tigers to protect their children.

As international pressure mounts and social media outrage goes viral, governments are forced to hire paramilitary units in an effort to stop the poaching and the killing. The results are heavily armed anti-poaching units, often run by former mercenaries and soldiered by poorly equipped and trained farmers. Many trigger-happy anti-poaching units patrol the bush with little accountability, and the results can be tragic. The losers are, inevitably, those who already have nothing. Inequality of wealth continues to drive a wedge between wildlife and communities.

Meanwhile, across the world a number of safeguards have been developed to try and keep animals and people apart. In Uganda and several other nations there are a number of experiments to keep elephants away from human settlements. Bee hives have been used as 'organic fences'. Elephants are fearful of bees, and the hives, with their occupants, create a 'barrier' through which elephants are reluctant to cross.

One of the other clever devices is an 'elephant repellent', a spicy mix of chilli and manure that is sprayed on the crops and smells and tastes deeply offensive to elephants but can be washed off by humans before the vegetable is consumed.

Elephants, cheetahs and lions have all been tagged, allowing researchers to track their movements and alert settlements when wildlife approaches. None of this is ideal but it is also the reality of trying to protect both the human and animal populations. But while the animal population suffers, the human population is booming. Uganda currently has a population of 45 million that will increase by nearly 15 million in the next ten years alone, creating massive pressures on the land. In fact, Africa's global population is projected to double by 2050 and to increase from the current 16 per cent of global population to 40 per cent by 2100. The pressures on the wilderness will be unimaginable, and the conflict will only continue or get worse.

All around the world, the wilderness is under threat by the expansion of farming and the sprawl of human beings. I could fill these pages with terrifying facts and figures but I don't want to depress you. While I would never want to shy away from the reality of the situation, this book is about the lessons we can learn.

The conflict created between humankind and nature is raw and ugly, but in the same way that we created the problem, we have the power to heal it. I recently spent time with a family of Extinction Rebellion activists in New Zealand. The family had retreated into the New Zealand bush to minimise their impact and to plan their actions. I have long admired Extinction Rebellion, but for the first time I got to spend a period in the wild with people who really, really care.

It was here that I first heard about the two bridges solution. The argument is that we can't build a new bridge without the old one. We could of course demolish the old one, but, by doing so, it makes building the new one much harder. Instead, you continue using the old one, but you are much more considerate. You use it less to preserve its fragility while the new one is built.

It was a light-bulb moment for me. There are some who are calling for a complete cessation of carbon-emitting activities – carbon-fuelled electricity generation, combustion-engine cars, gas boilers, aircraft. The whole carbon economy. The consequence? In my view, we would be thrown into the dark ages. While I don't doubt the urgent need for action, we also need to be considered in the way we unpick the spider's web of infrastructure on which the world currently relies. Blow up the only bridge we have and we are cutting off our nose to spite our face.

Extinction Rebellion are calling for governments to call a State of Emergency; their focus is on government action rather than individual action. But I beg to differ. I don't think we can afford to wait for governments to pass legislature. I think *we* need to be the tide of change. If you doubt that one small voice can make a difference, just look at Greta Thunberg.

If we all begin to champion change rather than trying to feel good about ourselves by picking up on others' failings, then we might actually get somewhere. How many times have I been criticised because I fly too much or because I wrote a book about the Land Rover and people therefore think that I don't care about the environment? I have thought long and hard about this. It would be far easier to abandon all opinion and step back from the debate, to avoid the charge of hypocrisy, but I do care. I have considered abandoning all air travel, but at what cost?

I make films about the impact of man on the flora and fauna. Without that 'bridge' I couldn't do my job or share the message.

It's inconvenient to admit, but my carbon footprint is larger than most, which is why I have tried to avoid lecturing or hectoring. Instead, I'd like to think that I can gently share what I have learned. It won't always be popular, but if there is one thing I have discovered over the years, it's that you can't appeal to all of the people all of the time.

The world is divided. So extreme is that division that it can be difficult to have a rational debate. Where once liberalism allowed for sensible reasoning, today people are unlikely to be swayed from their position. But we need to be open-minded. Sometimes you have to choose the least-worst option. We all need to make sacrifices.

As creatures of habit, we have become accustomed to the world that we have created. It is difficult to just turn it off. A little like writing this book. Seventy-thousand words is a scary prospect when you are staring at a blank page for weeks and weeks. Where do I begin? How do I start to consolidate those words into coherent thoughts?

The answer is of course by small steps. We are back to the expectation of haste. We demand instant gratification and instantaneous change, but we can't achieve that. We can't just press the reboot button. We can't switch life off and then back on again. We need to take small steps, then begin to lengthen them until they are a stride, a leap and finally a jump over that river, and that is the point at which we don't need either of those bridges.

How do we do that? We need to reduce our consumption to that of 'need' above 'want'. To reuse and recycle. To focus on happiness and health above money and wealth. To live for hikes and bikes rather than likes and swipes. To step out of the virtual world and back into the present one. To be compassionate and considerate and kind. Kindness sometimes seems to have been lost. I don't want to tar everyone with the same brush, but newspapers, television and social media have become increasingly mean-spirited. They look for faults, and are more likely to point out shortcomings than celebrate success. Where did encouragement and support go? The growth of extremism on both the left and the right has created a sense of isolation for many who are scared of the hate hurled at anyone who doesn't echo what each side wants to hear. I have lost track of the number of times people call me out for my own transgressions, so let me highlight here that I am not perfect, there is plenty of room for improvement, but I have tried and I will continue to try.

The wilderness is far less complicated and far less critical. A little like that forest bathing, I find she clears my mind and makes life feel less complicated. The growth of forest schools, wilderness retreats and forest bathing all around the world is also testament to her healing qualities.

Our relationship with the wilderness is as precious as it is precarious. Humans are excessive by nature. Inherently selfish and often unable to think beyond our own generation, we lack the empathy or foresight to consider changing the exploitative nature of that relationship. Those communities who rely on the flora and fauna around them, tend to be more sympathetic because they will experience those effects first hand, but for much of the world, shielded from the wilderness in big cities and towns, it is more abstract. Less relatable. The disconnect is at a tipping point, and if we continue to take more than we give, then we will be the losers.

At the heart of this climate debate is the wilderness. If we fail to heed her warnings, then we do so to the detriment of humanity, because although some people talk about us destroying the planet, the reality is that the world will far outlast us. It might become uninhabitable for humans, but the planet is not going to spontaneously combust or wither away. It is us, the human race and our future that is at risk, not the earth. We humans will be the ultimate losers. The survival of humankind is at stake. As Nelson Henderson once said, 'The true meaning of life is to plant trees, under whose shade you do not expect to sit.'

CHAPTER NINE

THE MOUNTAINS

'I'd rather be in the mountains thinking about God
than in a church thinking about mountains.'
John Muir

I have been scared and I have been frightened.

Fear is a basic primal emotion, but why do we fear? Fear is an integral part of our evolution and survival. Fear is often based on our instinctive drive for survival. The fear of snakes, insects, heights and closed spaces is probably based as much around self-preservation as it is anything else.

Fear can manifest itself in many ways. It is the basis for fight or flight. The adrenaline produced through fear can help us physically to fight, or it can be the warning sign to avoid danger. The problem comes when we no longer have any control over the 'fear'. Perceived fear is very different from actual fear.

The fear of clowns probably has more to do with watching too many Stephen King horror films than with the actual danger posed by clowns. Fear of heights seems a much more logical thing to be scared of. The problem is that if we listen to the fear, then we also begin to fear the fear, and when we fear the fear, we risk avoiding whatever it is that scares us in the first place.

The first time I remember real fear was when I got stuck on a mountain in Ecuador. I was 19 and I was living with a family in the capital, Quito. A couple of mornings each month, the clouds would clear and I would stare in awe at the soaring silhouette of Cotopaxi, one of the tallest active volcanoes on earth. Her snowy 6,000-metre peak would glisten pink as the sun rose in the Andes. She was beautiful and she was mesmerising.

Like a siren calling me, she led me on to dream about climbing to the summit. I was seduced by the romance and the promise of the views from the top. I was intoxicated by the idea of climbing. I was a mountaineering virgin, but that wasn't going to get in between me and my lofty ambitions.

I soon found myself on an expedition with my friends Guy Hedgecoe and Guy Edmunds to climb my first big mountain. To say that we were ill-prepared says more about my naivety than my laziness. We packed a rucksack full of peanut butter sandwiches and chocolate bars and a few days later we were at the mountain refuge on the snow line at 5,500 metres.

We were at base camp and we were already exhausted. Hungry and tired, Guy Edmunds developed altitude sickness and left Guy Hedgecoe and I to tackle the summit without him. We had hired a local guide, who, it transpired, had as little experience as us, and before long we found ourselves trapped on the side of the mountain in a huge storm. Without good leadership and against our better judgement we continued to climb, reaching the summit just as the storm enveloped the mountain.

Visibility was reduced to just a few feet, we were ill-equipped and ill-prepared and we still had a dangerous 500-metre descent back to the refuge. Just when I thought things couldn't get any worse, our guide started suffering from altitude sickness. We were stuck on a dangerous peak in the middle of the volatile Andes in a storm, and for the first time in my life I felt fear. Real, instinctive, makes-you-want-to-cry fear. I sobbed and wretched. I had lost control of the situation. For the first time in my life, I had no control of the situation. It was terrifying.

We all experience fear, but it is often fleeting. The slow dawning fear of being stuck on one of the highest active volcanoes on earth was entirely new and elicited a series of emotions I hadn't yet experienced. Fear can take you in two directions. You can either focus the mind, harness the adrenaline and take action, or you can surrender yourself to your circumstances.

You can never really predict how you are likely to react until you experience real fear for the first time. My initial reaction to curl into a ball and weep was soon overwhelmed by a rush of adrenaline. We soldiered on, making clear decisions as we reached crevasses and drops. Our guide was hallucinating and incoherent. I'm not really sure how, but we made it back to the refuge just as a rescue team had been assembled to come after us. I have rarely been so embarrassed, but we had overcome fear. This wouldn't be my last encounter with terror. Many such encounters would occur on wild, remote, volatile mountains around the world.

'It is not the mountain we conquer, but ourselves,' said one of the world's greatest mountaineers, Sir Edmund Hillary, who, with

Tenzing Norgay, was the first to reach the summit of Mount Everest. There is a great deal to be said about mountains of the mind. They represent lofty ambition, danger and ultimate achievement. But of course, mountaineering also comes with inherent risk and danger. You are at the whim and mercy of the mountain elements. Perhaps almost more than any other environment, mountains have a way of exposing your vulnerability that is difficult to avoid. There are so many variables that you are constantly challenged to adapt to new circumstances. Perhaps that is the draw? Perhaps it is the reason we are fascinated by mountains?

I don't think we take enough risks as a society. We have become risk-averse. We have tried to cushion and protect ourselves, or we ask the nanny state to protect us from risk entirely. But at what cost?

Humans are born with instinct. Our very survival depends on it. You only need to observe the animal kingdom to get a sense of how strong our survival instinct can be. Any interaction with the wilderness comes with an element of risk. Changeable weather, lack of shelter, vulnerable landscape, exposure, lack of food and water, wildlife, poisonous flora … the list is endless, but compared to the comparative safety of the 'civilised' world, any interaction with the wilderness involves weighing up risks – and nowhere is this more visceral than in the mountains. By reducing our exposure to any risk we fail to build any defences. We never make mistakes, so how can we learn from them?

The thing is, without risk we can't live, learn or grow. We need to let risk into our lives to achieve and succeed, otherwise it is no different to the flatline of happiness. Each of us takes different risks in different fields and we rise or fall accordingly. Some of us will risk more than others financially, while others will take academic risks or relationship risks or expedition risks. To achieve success you need to take a risk otherwise it isn't really success. The word itself implies that you have done what others have failed to do, and to do that requires facing risk head-on.

If we all sail through life with ease and comfort then we might avoid suffering and hardships, but we also fail to grow. Life becomes a homogeneous, vanilla flatline of mediocrity. We need bumps along the way. We need a defibrillator to shock ourselves out of that flatline. Risk is an essential part of development; by shielding ourselves we inhibit growth.

Risk has become an important part of my life. Risk has helped me grow and strengthened my confidence and spirit. The constant risk of failure has become a driving force. Perhaps it stemmed from those early failures. The reality is that risk has helped me exorcise my demons and rebuild my confidence. Of course with risk comes the likelihood of failure, but maybe that's the point? The greater the risk, the bigger the rewards.

Risks have always been a part of humanity. Right back to prehistoric man, risks would have been taken in the pursuit of reward. Do I take on this giant woolly mammoth with my spear? Or should I go for the smaller deer? The rewards were obvious, as were the risks.

Taking risks is as important in business, economics and work as it is in sport, relationships and friendships. We are constantly taking micro-risks as we gamble with everyday life. Should I risk taking the bus or the tube? Should I invite the person I really fancy out for a drink? Should I tell my partner about those new shoes I just bought? Should I buy that new phone or wait until the next one comes out? These really are micro-risks, tiny in the grand scheme of things, but risks nonetheless.

If you invest your money, you could go with the solid high-street bank or some anonymous online bitcoin type of investment. If you are buying shares you might go for Disney, in which your money is likely to be safe but the returns limited, or you could invest in renewable energy in Papua New Guinea, a far riskier proposition with potentially higher returns. Whether we notice or not, we are constantly weighing up the pros and cons of everything we do. Calculating, comparing and contrasting the decisions we make. It's one of the reasons I still hate decision-making. It's so much easier to defer to someone else, but at what cost? It is detrimental to our self-esteem and to our feeling of self-worth.

Risks vary in scale from the mundane to the life-changing and even life-threatening. Deciding whether you are going to risk going out without a raincoat on an April day in Manchester is a very different kind of risk to leaving your tent in Antarctica without a coat to take a pee in the middle of the night. The former will leave you wet. The latter will probably leave you dead.

Comfort and safety are nice, but nothing ever grows there. A ship is safer in a port than it is on the open ocean, but ships are made for the sea. Aeroplanes are safer on the runway than in the sky, but planes are made to fly. We humans are given one life. It is up to us to decide how we choose to live it. We could wrap ourselves in cotton wool and shield ourselves from all risk, but at what cost to actually living our lives? Is life something to be enjoyed or endured?

You have to ask the question why we are here on this earth. We certainly aren't here to make a positive contribution for future generations. Our disregard for the planet and exploitation of Mother Nature is testament to that. So if we aren't here for others, then what is our purpose?

It is so easy to sleepwalk through life without question. Eat, sleep, work. Eat, sleep, work. In urbanised society there are so many distractions that we hardly have time to keep up with friends, let alone ask why we are here. But take away the distractions and the background noise and what you are left with is a sense of deep contemplation.

Mountains are spiritual places that have the power to overwhelm, and part of this is because the climb leaves you little room for distractions. Up high, the magnitude and wonder of the wilderness almost screams out at you. During those few months on Mount Everest, stripped of interactions and distractions, my mind became hyper-sensitive and beset by questions, booming from the sky like the voice of God: 'Who are you?', 'Why are you here?', 'What's it all about?', 'What is the meaning of life?'. I know it sounds Monty Pythonesque, but there really is a powerful contemplation that fills the vacuum vacated by everyday distractions, to the extent that I was constantly questioning why I was here and what my purpose was.

* * *

The first time I really challenged myself was when I was 16. Pretty late in life, I agree. Up to that point, I had been debilitated by shyness, so I decided to confront it head on, by doing a comedy routine in front of the whole school, on my own. Now let's just put this into context: I was very shy, I was embarrassed about my spots and my big nose, and I lacked self-esteem. I was one of the odd kids. I was an introvert who just didn't fit in. And yet somehow I decided to take a risk, a massive risk, and stand up in front of a thousand pupils and teachers and deliver a comedy routine I'd written myself.

No one asked me to do it. No one suggested that I do it. Indeed, assemblies were usually driven by teachers, not pupils. As an angst-filled teenager the risk couldn't have been higher. It was off the scale. The risk of long-term damage to my self-esteem cannot be overestimated. But I did it. I strode onto that stage and I unleashed my carefully rehearsed routine onto an unsuspecting school. I am embarrassed to think about it now, more than 30 years later, and I still blush to see myself there. But in that moment, in the spotlight, it felt as if I was wearing a suit of armour to protect me.

In all honesty, I don't recall much. I don't recall laughter. I don't recall clapping, but neither do I recall booing nor soggy tomatoes hurled onto the stage. But I 'felt' such a huge sense of satisfaction and my 'inner' peace gave me an outer confidence that assumed it had gone down well. 'Well' is probably the wrong word, but there was an unspoken 'respect' that I had risked so much to stand on that stage alone. I had laid myself and my vulnerabilities bare for all the world to see, and while the reception and reaction had been muted, if not silent, I knew that there was a shared sense of admiration that I'd had the balls and the chutzpah to do it in the first place.

The risk had paid off. My confidence had grown as a result. It was like I was testing myself. Testing my own resolve. Exploring my own boundaries. I doubt if anyone who was in that assembly would ever even remember the routine. They certainly would have had no idea of the effect it would have on my own relationship with risk.

Ever since that time in school I have embraced rather than eschewed risk. Now don't get me wrong, I am not going to sit here and encourage you all to go and risk climbing K2 or kayaking the Pacific Ocean, but I would counsel you sometimes to take yourself outside of your comfort zone. Of course the risk should be sensible. I would like to think that the risks I take are calculated risks. I try to set myself parameters and guidelines. I respect the environment and its flora and fauna, but I take risks nonetheless.

But when I became a father ten years ago, I felt a paternal draw towards safety. I realised that to risk your own life, health or well-being impacts directly on your loved ones. So does that mean that risk can never be inclusive, or that it can ever be considered a selfless act? I've often wondered about this and the answer is complicated. I would like to think that, by embracing risk, I am investing in myself and those around me, by staying engaged with and excited by life. Risk bolsters inner confidence and happiness. There is that word again: happiness. That doesn't mean I am unhappy when I am not taking risks, but it means I am not being true to myself.

* * *

I've taken a while to get to the riskiest expedition of my life, but for me it was a trip that summed up the very essence of risk.

Climbing Mount Everest was neither the hardest nor the most arduous of my challenges, but it was without doubt the most dangerous. Everest had been a lifelong ambition. Ever since I was a small boy I had marvelled at the heroics of those early climbers. The mountain became an obsession. I read every single book on the subject and I was open-mouthed in admiration at the bravery of those early pioneering mountaineers who became my role models. As a child, though, Everest was only ever a dream; she was a mountain for courageous sportsmen and women willing to risk their lives to reach the highest point on earth.

Thirty-five years later, as a father, I found myself conflicted. On the one hand I was encouraging my children to follow their own path. To follow their own dreams and pursue happiness, while at the same time I had abandoned my own dream of one day climbing Everest. I was torn between the selfish pursuit of aspirations over the selflessness of parenthood in which we quite reasonably make sacrifices for the benefit of our own children.

Becoming a father is the greatest thing that has ever happened to me. My children, Ludo and Iona, are my world. They are my oxygen, my shadow and my heart. My children are a part of who I am; they are of course independent little beings, but my own happiness is entirely controlled by theirs. As the old saying goes, you are only ever as happy as your unhappiest child.

I love being a father. The opportunity to teach, share, nurture and provide. It tests us and it rewards us. I wouldn't have it any other way, but it also forces us to make decisions. We have to decide what is right and what is wrong, not just when it comes to discipline and the lessons of life but also when it comes to ourselves.

Which brings me back to the question at hand: is it ever okay for a father or mother to climb Mount Everest? Perhaps we should refine that: is it ever fair for a parent to risk his or her life? I suppose it depends on both the scale of that risk and also on the returns. A little like a risky investment, it depends on what you stand to gain, over what you stand to lose. For me, climbing Everest was as much about what my family stood to gain as it was about my own personal growth. It's a convenient sentiment, I agree, but I think that life is as much about the influence of our own morale and happiness on others. Happiness can be infectious. If you spend time in the company of happy folk, more often than not you find your own spirits lifted and your mood lightened. I love being around happy people for that very reason. There is nothing worse than being stuck with a glass-half-empty pessimist. I don't have time for doubters and haters. I have always lived life to the full. I have tried to seize every opportunity and enjoy the ride.

Climbing Everest was an opportunity to use all the lessons I've shared with you in this book and to push my own limits. How often do we reach a point where we give up? We find an excuse why we shouldn't do something. Sometimes there really are a thousand reasons why you shouldn't do something and only one reason why you should, but that is the difference between success and failure. That is the difference between those who stand on the stage and those who hide in the wings. Do you step into the sunlight or do you hide in the shadows?

I don't want to repeat my whole Everest story; I wrote a book about that, but, once again, I think it's important to recap parts of that story in order to share with you the lessons I learned from the experience.

One underlying principle for the whole trip was 'respect'. I was conscious of the importance of respecting the mountain, the mountaineering community and the local Sherpa people of Nepal. I wanted to honour that respect by preparing for the climb as best I could. Although I had climbed a number of peaks over the years, they topped out at 6,500 metres. Everest, at 8,850 metres, would be the highest by a long way. I asked my old friend, Kenton Cool, veteran of 12 summit climbs, if he could help me with my climb. Together we set a two-year training regime that involved climbing multiple mountains all around the world, from the Bolivian Andes to the Nepalese Himalayas and the French Alps.

Although Kenton would join me on the climb, I wanted to experience Everest with another teammate, someone who would share the fear and the risk and the excitement of the journey. You may be surprised to hear this, but I am quite a lazy individual. I need to be pushed. It was one of the reasons that James Cracknell and I had such a successful adventuring partnership. James was always able to get the most out of me, and I wanted to repeat that on Everest. Although Kenton had the experience to do the same, this also counted against him. He knew that mountain better than almost anyone, and the risk was that I would absolve all sense of responsibility to him. It is so easy, especially for lazy folk like myself, to defer to other people, and that wasn't the object of this journey. I wanted to experience the mountain for myself, along with all the tangible risks involved in climbing it.

I didn't know former Olympic gold medal cyclist Victoria Pendleton. I had always admired her from afar: her fierce, competitive spirit and drive, but most significantly her individuality. Many athletes, particularly Olympians, become institutionalised. They spend so long working to a routine, being told when to train, when to eat and when to sleep, that they begin to lose their personality to sport. I'll probably offend a few sportsmen and women with that observation, but in my experience it is true. I think it's one of the sacrifices they make to their sport. When you want something enough, you will do whatever it takes, but Victoria had always appeared to be true to herself. She didn't suffer fools. She spoke out and she danced to her own beat. I liked that. In some ways she reminded me of James. A hard taskmaster, hard on other people, and even harder on herself.

I wanted someone to get the best out of me, but also to share the burden of fear, because Everest is still a dangerous mountain. What you end up experiencing is usually very different to what you imagined. Perception and actuality are diluted by the power of the mind. It's important to discuss perceived danger and actual danger. Mystery and the unknown can be an unbearable burden. You may be surprised to hear it, but I don't like change and I'm terrified of the unknown. I rather like conformity and consistency.

In part, the fear of the unknown comes from the power of the mind and the imagination. By and large most challenges, expeditions and journeys sound more impressive than they actually are. The power of the mind to build a hellish nightmare out of a perfectly feasible task should not be underestimated. When I imagine someone on Mount Everest, the image built in my mind is far more scary than the reality. That is not to diminish the dangers or the achievement, but it's also why it can often be harder for those left at home than for those who are there. The man or woman on the mountain has the benefit of understanding the actual place, while those back home are forced to control the visions of their imagination, or of their nightmares.

Add to this another complication. For some reason, we have a fascination with disaster and near-death experiences. Most of the popular travel and adventure books and films rely on disaster, failure and death as a narrative. The press love to fixate on the failed expeditions as much as they do on the successful ones. For every reported death on a mountain, the papers fail to report on the hundreds of successful summits. For each capsized sail boat at sea, we rarely hear of the successful ocean traverse of thousands of others.

More often than not, we tend to focus on the tragi-heroics of those who were lost in avalanches or in storms at sea. And if the stories don't end in death, they will focus on near-death, the stories of those left for dead dangling from mountains or marooned at sea on a life raft. It isn't really that surprising that we have a fascination with stories of survival. It plays into our own primal fears; we project ourselves into those circumstances and we begin to use our overactive imagination to 'experience' it ourselves. We, the journey folk, are fully culpable for exaggerating this narrative. The story of a completely successful journey is not what people want to hear. They don't want to read about how perfect it was. They don't want to watch a film of a clinically executed expedition.

People want to experience life and near-death vicariously. They want to frighten themselves through other people's experiences. I get it. I've done it myself. But the fallout can be quite dramatic, and I suspect I am part of the problem when it comes to the alienation of the wilderness. When I say 'alienation', I mean that there is a notion that the wilderness in general is a hostile environment that wants to injure or kill us.

The notion of man versus the wild is not entirely true. I think in general we have a fear of the wilderness. We like to control our environment. When you consider a modern urban environment, we have taken full control of that landscape. We coat it with tarmac and concrete, we gate the parkland and manicure the flora and fauna. We look to create a pristine environment of which we are in control. Of course, even this is only an ideal. It takes an earthquake, a flood, a wildfire, a tornado or a hurricane to remind us of our vulnerability, particularly within the towns and cities. For many, the wilderness, in her many different forms, is an unknown, challenging place in which we are forced to battle against nature for survival, but this is nonsense, and, deep down, I think we all know it. We have already 'othered' the wilderness through popular culture. We have created an environment of conflict: a hostile place to fear rather than admire. It might be hard to accept, but it's important to understand what the perception of Everest is to my loved ones.

For me, Everest was a beautiful mountain waiting to be climbed, but to my family and friends it was a dangerous mountain that has claimed hundreds of lives. It's so easy to dwell on the negatives rather than the positives, and therein lies half the battle with risk. Focus on the positives as well as the negatives, and that risk becomes more manageable.

How do I evaluate the risks I am willing to take? I don't have a formula but it's important to be open-minded. You need to be sensible and sensitive with the decision-making. So here is how I rationalised Everest. I began with the negatives.

It's dangerous. It has claimed many hundreds of lives. It will involve many months away from my loved ones. The risk of failure is very high, and summiting is not guaranteed.

And then I looked at the positives.

Thousands have successfully climbed to her summit. Far more live than die. If it was easy, everyone would climb it. Risks are an important part of life. With teamwork and experience we could mitigate some of those risks.

I simply out-argued the negatives. The risk of death will always deafen the argument. It's hard to rise above the power of that word. There was a time, my wife Marina reminded me, when one in ten people perished on the mountain. That's a 10 per cent chance of death. That is the same return on a healthy financial investment. Ten per cent? Is any challenge worth a 10 per cent risk of death? That's a hard one to justify, particularly when you have a family of your own.

The reality of course is that it isn't really 10 per cent. There were terrible years of mountain cowboys dragging inexperienced climbers onto the mountain, and storms and earthquakes that have skewed the statistics. Newspapers like to repeat the figure, but if you go beyond the statistics and try to interpret the reasons for those deaths, then you begin to dig beneath the sensationalism and start to reach the truth. For me, Everest was always so much more than the baggage of her news headlines, which were heavily skewed towards disaster. I just had to persuade my family of the same.

By now I had teamed up with Victoria Pendleton. She had retired from cycling and was looking for another challenge. I asked if she wanted to join me to climb Mount Everest and she agreed.

And therein lay the root to many of her eventual difficulties on Mount Everest. I asked her. Until that moment in her life, I don't think she had really considered climbing Everest, and she didn't have the burning childhood dream. She didn't lie awake at night visualising the climb and dreaming of standing on the summit. For Victoria, Everest became the next Olympics, a challenge to rise to. A challenge to work towards in the spirt of excellence and competition.

Kenton set us a series of mountains to climb. The first was a three-week expedition in Bolivia, culminating in the 6,500-metre Illimani, a brute of a mountain. I was nervous. I'd had mixed results on previous mountains. Altitude sickness had struck me while climbing the 6,000-metre Cotopaxi in Ecuador and I had failed to get a team to the summit of Chimborazo during the BBC series, *Extreme Dreams*. The odds were stacked against me, but I knew this was my one chance. Waver and I'd lose the battle.

Marina and my family had made it quite clear that this would be my one and only opportunity to try and climb Everest. Any failures along the way and it was over. Those weren't their rules, they were the unwritten contract between a family.

I threw myself into that preparatory expedition in Bolivia. I went with the utter conviction that I would storm through it. I never doubted myself. Sounds simple, but believe me, my life has been full of self-doubt. It's called imposter syndrome, the belief that you've somehow blagged your way accidentally or taken someone else's life. On the Atlantic, in the desert, in Antarctica, I was always plagued by the deafening voice of doubt, but for some reason, in Bolivia, my mind was silent. And I was happy. So happy. Those mountains sang to me. They called to me. They were so beautiful. I could visualise myself from above as I climbed ever higher. It was hard. It stretched me physically, but there was such beauty and peace on those mountains.

From the beginning, though, Victoria was plagued by doubt. She never told me this, but I could see it. It feels slightly dishonourable to be talking about my friend like this, but I think in the spirit of the book it's as important to share the errors as well as the successes. You see, I don't think Victoria ever really believed in herself. I don't know why, because I have rarely met an individual so determined to succeed. The problem was the unknown. The velodrome in which she had won world championships and Olympic gold was her environment. It was her comfort zone. Get her on that track and she owned the place. There was never doubt. Her diet, her training, her bicycle were so finely tuned that all she needed was confidence and the medal was hers.

Mountains are very different beasts. Unpredictable and mysterious. It is those qualities that attract mountaineers. And those qualities that make them so dangerous. I can remember staring at the vertiginous slopes of Illimani and her vast snowy summit and struggling to comprehend how on earth we were meant to climb all the way to the top. She looked impossible. Really. I couldn't comprehend how we'd ever do it.

And that's the beginning of the battle of mind over matter. I knew we had the physicality needed to climb her and I didn't doubt that we had the ability. We certainly had the expertise in

Kenton Cool, but none of that matters if you don't believe that you are capable in the first place. And to believe, you have to want.

Now let me share with you another unspoken thing that happens on expeditions. I suspect it happens in sport all the time and I don't doubt that it is also part of the business world.

It is the strength and weakness model of expeditions, in which, as one person weakens, another strengthens to fill the vacuum. It's quite an ugly sentiment, if I'm honest, a little like parasites feeding off our blood in the sense of taking strength from another's failings.

I imagine it happening on the sports track: as one athlete accelerates, the person behind them either rises to the challenge or they fall even further back, at which point the person at the front receives an added burst of adrenaline and surges ever further forward, creating an even bigger gap. It's a vicious circle. I have experienced the same on many expeditions and journeys. When one team member is low, the others tend to be high, even higher than normal. It's as if they are feeding off the negativity to feel an exaggerated positivity.

It sounds cruel but it's almost as though the others begin to feel even better about themselves. They almost find a sustenance from the misery. As I say, it's ugly but it's true. I would never wish the hardships of expedition misery on anyone, but where Victoria struggled, I strengthened. If I could have shared some of that strength I would have done so, but once a chink in your armour appears at altitude, it is a slippery slope.

Victoria made it to the summit of Illimani, but she was wracked by self-doubt all the way up and she beat herself up all the way down. Although we had summited together, in good time, she felt she had failed. I was and still am so confused. How had she failed? She found it difficult to communicate her disappointment, and although we had successfully completed the Bolivia expedition, Victoria felt she had failed, and it would be a difficult disappointment to come back from.

A few months later we were back in the mountains, this time the Himalayas and a famous mountain called Island Peak. She was a smaller mountain, just 6,000 metres, but Kenton wanted to introduce us to the region in which we would return for our Everest climb. Once again, Victoria was crippled by inner doubt. I knew she was capable, Kenton knew she was capable, but I'm not sure Victoria really believed it. She gave up just a few hundred feet from the summit. I was more disappointed than she was. Maybe, inadvertently, Victoria's doubt helped me rise to the challenge. Although Kenton was our guide, paid to help us achieve our goal, I also didn't want to let him down.

By the time we reached Everest, I think our cards had already been played. While Victoria arrived brimming with confidence and positivity, the war of attrition eventually began to take its toll and after just a few weeks on the mountain she suffered extreme altitude sickness and abandoned her attempt. In many ways, life became easier after Victoria withdrew. I missed her, but I was able to focus on the mountain once again rather than Victoria.

Everest is so much more than a physical challenge; a little like the Atlantic Ocean, it is very much a test of psychological strength. The combination of the unknown, the dangers of the mountain, the unpredictable weather and the tragic stories of death create a pretty ominous atmosphere. I shall never forget the memorial on the way to base camp, hundreds of stone cairns and plaques for those climbers who had never returned from the mountain. It is a wretched place, where a strong wind sweeps across the plain. It is cold and inhospitable, ugly and deeply, deeply sad.

On reflection, I think it was important to confront our mortality through those who had lost their own lives in the pursuit of their dreams. There was no difference between me and those hundreds of others who had died. Bad luck and misfortune were all that separated me from them. It was haunting and incredibly emotional walking around that place of hope and loss.

The whole Everest experience was both beautiful and awful. The fine line between life and death is everywhere. You can taste it, smell it and see it. Not just in the sheer drops that tumble thousands of feet just inches from your feet, but in the dead bodies on the mountain. The bodies of other climbers, all 300 of them, who died in the pursuit of the same goal. I'd been arrogant to assume there was any difference between my ability and theirs. It was in the hand of the gods, like Russian roulette. The mountain was either going to let me live or I would die. Summit or fail. Mother Nature was very much in charge.

This is one of the reasons that mountains are such levellers. We abstain from control. In society we are used to controlling our environment, but on mountains the environment controls us. No amount of risk management could ever account for avalanches, storms, crevasses, icefalls or human error. If I'm honest, I was surprised by the dangers of Everest; perhaps I had been so good at controlling my mind and the inner voice of doubt that I had visualised an over-sanitised environment, free from danger.

To summit Everest requires a series of rotations ever higher up the mountain. Beginning at base camp, you start by walking through the notorious Khumbu Icefall to Camp 1 and then return. You recover, and then a few days later you begin another rotation on the mountain, this time up to Camp 2 at 6,500 metres, the same height as our training mountain, Illimani, in Bolivia. You then spend a few days acclimatising before descending, recovering and then starting another rotation. You do this until your body has adapted to the ever thinner air, and you then plan your summit bid, which consists of one circuitous ascent and descent that usually takes five days.

For many, the most feared obstacle is the Khumbu Icefall, a notorious 1,000-metre ascent through jagged blocks of ice, across yawing crevasses and beneath towering seracs that can collapse without warning into the icefall, tearing it apart. Many, many people have lost their lives in this icy labyrinth over the years, and for me it was without doubt the hardest part of the climb, not in a physical sense but because of having to overcome my own fear. The first time I saw the icefall from Base Camp I laughed out loud. It looked ridiculous. Impossible. I genuinely couldn't even fathom how we could navigate through it. I could just make out the pinprick of a dozen climbers, high up in the icefall, slowly descending back to base camp. I would sit in the porch of my tent and stare at those pinpricks as they appeared and disappeared across the icy obstacles. I tried to visualise myself in there, but I had no context. I had never experienced an icefall at the base of a glacier and the result was that I created an ogre in my mind. One that beat me to death.

It was dark the first time we ascended the icefall. With my head torch illuminating a small patch of snow and ice ahead of me, the wider landscape remained invisible. I had been dreading the infamous ladders strung out across crevasses. They would bend and bow as each climber slowly stepped out over the seemingly bottomless cracks in the ice. Fall and you might tumble 1,000 metres to an icy tomb below. Heart in mouth, I slowly stepped out onto the first of the ladders. The claws of my crampons wedged themselves into the rungs of the ladder, threatening to tip me off my balance with each step as I tugged at my boots for freedom. Up, over, across, down, around. It was an icy version of Go Ape, only here, if you fell, you'd probably die.

Until now, the landscape had been abstract, cloaked in darkness, but as the sun began to rise over the towering Himalayan mountains, I could finally see the labyrinth beyond. Both terrible and beautiful, as the sun cast a pink hue across the landscape, I could see base camp not far below. We had come far but we had farther to go. As the sun rose, so too did the inner voice of doubt. Quietly at first, but ascending in volume to a deafening roar.

As we neared the top of the icefall I noticed a change in Kenton's behaviour too. Gone was the relaxed, calm Cool, replaced by a much curter version. 'Don't stop,' he warned, 'I want you to walk through here as fast as you can.' He spoke with urgency. We both looked up at the towering blocks of ice that teetered above us. These were the origins of avalanches. One shift in the snow and ice above and they would collapse into the icefall.

No amount of experience or planning could anticipate an avalanche. It was entirely down to luck. Now this is where my whole risk model falls apart, and is replaced by fear. I have known fear in my life before. I experienced fear in Antarctica when we found ourselves alone, un-roped on a crevasse field, and I had experienced fear while diving with crocodiles. But it had only ever been fleeting. Suddenly here, on Everest, fear knotted my stomach and froze my limbs. My heart raced and, despite the cold, I began to sweat. I could hear buzzing in my ears as I soldiered on, not at any haste as my body was constricted by the gradient and the thin air. I found myself glancing up at the ice above. Occasionally, the roar of a faraway avalanche echoed across the icefall. I just hoped it wouldn't be here.

Part of mountaineering psychology requires finding an inner peace and contentment. On the mountain it is essential that you are in 'the now'. You cannot let your mind wander. Bring in outside concerns and worries and you are screwed. I have experienced this in the past, but never was it more real than on Everest. The clarity of thought required to navigate each obstacle was exhausting. And all along you had to remember that what goes up must come down. We would be forced to navigate through the Khumbu Icefall at least six times, maybe eight.

I would lie awake in my tent at night, listening to the groaning of the glacier beneath and the booming echo of avalanches in the valley beyond. 'You shouldn't be here,' my brain would tease me. 'You aren't qualified,' she would continue. 'SELFISH, SELFISH, SELFISH,' she would chant. The brain sometimes feels like an enemy, out to jeopardise our chances or jinx our opportunity. To counter what my brain was thinking, I tried to revert to the positive. 'You have one shot at this,' I repeated to myself over and over. It was exhausting.

Many of those who have perished on Everest have succumbed to what is known as summit fever. It is the delirium that exaggerates the importance of summiting against all better judgement. The mind becomes so obsessed by the notion of reaching the summit that it persuades the climber to carry on despite the incalculable risk to life.

Getting to the summit is only half the journey. You need oxygen in the bottle for the descent, and it is the return journey that claims many lives. People who have overexerted themselves on the ascent have left no reserves in their body for the return journey. I had promised to my family that I would not be seduced by the siren of the summit. That I would be sensible and not look too far ahead.

Each day on the mountain brought me ever closer to that summit bid, until three weeks had passed and Kenton announced that we were ready for the top. So it was through the Khumbu Icefall, past Camp 1 and into Camp 2, where we rested for the night. A game of cards to settle the nerves and a cup of tea to settle the nausea and then it was onwards and upwards, this time to Camp 3. We pressed on through a gathering storm, reaching Camp 4 at 8,000 metres, where by now we were higher than I had ever been. We were in the death zone. Here we would only get one in three breaths of air. Life is not meant to survive at this altitude, and the stopwatch had already started.

A storm raged. Tents were torn. Climbers had been lost. It was terrifying. Rarely have I felt so outside my comfort zone. Here of course we were beyond rescue. Survival relied on sensible decision-making and sensible risk-taking. This was where having Kenton's experience really paid off.

Dawn heralded a break in the weather. Suddenly, the wildly inhospitable Camp 4 seemed oddly safe. I read a poem out loud.

In the early evening we set off to climb the final 850 metres. Head down, small steps. Trying to control my breathing. I had a fear of being overcome by panic. I had to control my emotions. At 8,500 metres, my oxygen bottle exploded. The valve sheered and I lost all of my oxygen and the use of the mask itself. It was like that capsize in the Atlantic all over again. Heroically our Sherpa, Ming Dorjee, gave me his bottle and mask and he returned to Camp 4 where he would find another bottle and mask.

By now the sun had begun to rise. First a thin pink line on the horizon, then an orange glow. With the first glimpses of daylight came the reality of our isolation.

BANG.

Our cameraman Mark's oxygen bottle and mask also blew. Kenton urged me to carry on. Alone. Higher and higher I climbed. There was no one ahead of us and Mark and Kenton had blocked the route below while they sorted a replacement bottle and mask. And before I had time to think, I found myself marooned on a near vertical slope on my own. I couldn't go up and I couldn't go down. I dug my crampons into the ice slope and looked around.

The sun broke the horizon and a shaft of light illuminated the ice and slowly warmed my face. It was still bitterly cold, but in that moment it was like the world had stopped. I had never known such beauty. I had never experienced such solitude and isolation. This was the reason I was here. Not to stride magnificently onto the summit, but for the journey. The climb itself.

So many of us get caught out by this. We become so focused on our goals and objectives that all the rest becomes arduous. The journey becomes a burden. 'Don't wish away your life,' my grandmother used to say. How many of us have endured the trials of a tedious task, wishing away that moment for better times? The problem is that you never know what is around the corner. Too many of us are blinded by that destination, the metaphorical summit. But here, at that moment, on the side of that mountain, alone, I had never felt such happiness. I was terrified, but that didn't matter because the beauty of that moment transcended anything else I had ever experienced. It was more than a single moment but a culmination of events throughout my life.

As I stared at that vast expanse of mountains and glaciers stretching as far as the eye could see, I was reminded not just of my mortality but of the beauty of this life. The beauty of this world of ours. I was looking down on an untouched wilderness. My moment of 'isolationism' didn't last long. Another oxygen bottle shuffle and the climb was on once again. Ahead of me I could see a small saddle, over which I felt certain was the summit.

Here's some advice I should have remembered from long car journeys during my childhood. Never ask how much further. The answer is bound to disappoint.

'How much further?' I asked Kenton as we crested the brow of a false summit. 'About two hours,' he replied matter-of-factly. Now, two hours in the grand scheme of things is nothing. You could watch a film in that time, but here and now, after four weeks of suffering, it felt like a lifetime.

Head down, I soldiered on. Life is hard in the death zone. Perhaps it is our vulnerability that slows us down, but nothing seemed real. I was now at 8,800 metres, at the famous Hillary Step. I was 50 metres from the top of the world. Finally, for the first time, I really believed it was possible. I thought that I really might make it.

BANG.

Another oxygen bottle gone. My chance to carry on hung in the balance. We had no more spare bottles or masks. The Grim Reaper had appeared once again. I began to panic. Not at the missed opportunity to reach the top but at the very real fear of not getting down. History had documented the swift decline in climbers' ability to stay warm, focused and alive without supplementary oxygen.

Swiftly, Kenton removed his mask and placed it over my face. He would wait for other climbers from the Chinese side, in the hope that he could somehow make or borrow a spare mask. It was a gamble, a huge risk, but also a calculated one that was based on 12 previous successful summits. Those final few metres were the sweetest of my life. Ten metres, eight, five, three, two …

And then I was there. Standing on the summit of Mount Everest, at 8,850 metres. The highest point on earth. There was no euphoria or fist-pumping, champagne-cork-popping celebration. I looked out at the horizon and a sky so blue it seemed almost black. I could see the curvature of the earth, and hundreds of miles of wilderness. I have never known such isolation and beauty. I felt happier than I had ever been. The happiness was not based solely on the satisfaction of achievement, but a large part of it came from being in such a special, spiritual place.

Throughout this book I have often alluded to spirituality. I am not a religious person, but that doesn't mean I don't believe. Not necessarily in a higher being but in something. I suppose the wilderness is my temple, church and mosque. The wilderness with her trees, rivers, lakes, oceans, jungle, mountains, desert and savannah is a place that has taught me how to be happy. How to live my best life.

There are so many lessons we can take from nature. Some of them are obvious, others are a little more nuanced, but at the heart of it is our ability to be better people.

Everest was a sum of her parts. It really was so much more than just a climb, and although standing on her summit was a life-changing moment for me, it really was the journey and not the destination that affected me most. Maybe I got lucky. Or maybe it was my destiny all along.

The mountains are ethereal places that have existed in our minds for centuries. Mountains dominate our landscapes, calling all those intrepid, adventurous souls with their siren song. For me, mountains will always hold a great romance. They strip us of our egos and they test our resolve.

Which brings us back to why we climb mountains and what we can learn from our ascents. Mountains remind us of our mortality. Fitness, experience and even knowledge are no barrier against the changeable mountain weather, the lightning strike of altitude sickness and the unpredictable avalanches. Sir Edmund Hillary himself famously said that 'it is not the mountain we conquer but ourselves'. Mountains create a vacuum within, from which a deep and intuitive sense of searching begins. I have rarely felt such distant clarity. You are there but not there. It is almost as if you are looking down on yourself.

Mountains have long held a spiritual draw for those who want to purge and cleanse their bodies and their minds and search for a deeper meaning. Mountains force us to confront realities that many us try to avoid. I never went searching for answers but I found a lot of questions. I found risk and love and loss. Mountains are a little like great loves, in that to really love, you must risk loss. To live, you must take risks and face fear. Until you have confronted your mortality, you have never really known what it means to live. Mountains strip you of your ego and highlight the nakedness of our vulnerability. As a mountaineering friend once told me, 'Climb the mountain so that you can see the world, not so that the world can see you.'

RE-WILDER

'I am the wilderness lost in man.'
Mervyn Peake

I have been re-wilded.

Sometimes you don't realise what you have until it's gone. The wilderness has helped guide and reshape me.

I have learned so many lessons about myself and about others in nature's theatre. There is of course a striking difference between self-imposed risk and struggle and the unavoidable struggle experienced by many millions of people around the world. The knowledge that you will eventually return 'home' and back to normal life is very different to those stuck in poverty, misery or danger.

Running a marathon in front of adoring crowds that line the streets, at the same time raising money for charity, is a whole different type of suffering than the kind endured by a Kenyan child who must walk five or six miles to school and back each day.

Running out of food in the middle of the Atlantic Ocean is very different to having a failed crop or losing your job and being unable to afford to eat. They are both metaphorical holes you can find yourself in, but the former has an escape ladder of sorts while the latter can be inescapable for many. While I have experienced many emotions on my adventures, I would say I have experienced more of the physical effects than the psychological ones.

We live in a culture of blame. We like to rely on others so that we can excuse our situation by blaming them and not ourselves. People manage to get themselves into a foaming rage by accusing others of causing their woes.

In some ways it is human nature to find a common enemy. Rival companies, rival suitors, governments, businesses. If you read the news, you will see how the newspapers play the game by focusing on a particular group and building up national hatred. The result has been that everything has now become politicised. The economy, the health service, the environment, schooling, Covid-19. It never ceases to amaze me how politically charged almost everything now is.

When we lived on Taransay, it didn't take long for us to find a common enemy in the BBC and Lion Television who were making the show. We became wary and distrustful, which is odd because the only reason we were there in the first place was because they had invited us. To become distrustful of your host is strange and perhaps also slightly disingenuous.

On Everest we became distrustful of other climbers. Worried that they might jeopardise our climb by spreading germs, we washed our hands furiously after each interaction. Many wore face masks and the majority 'quarantined' and closed their camps off to any others. It was a 'lockdown' before anyone had even heard of Covid-19. On the Atlantic Ocean we became wary of the race organisers.

Even in the wilderness with the wild folk around the world, I have found that they have a common enemy, that enemy being 'the system'; in other words, wider society and consumerism.

Suffering and risk through choice means you absolve anyone else of blame and embrace full responsibility. By proactively enduring, suffering and risking, you are experiencing a very different kind of pressure. To knowingly, and willingly, take yourself away from the safety of your comfort zone, you are choosing to use risk as a template for growth.

There are plenty of times in life when we might unexpectedly suffer hardship, danger and hunger. Those experiences toughen us and strengthen our resolve. I'm not saying they are good, but they certainly build character. When, like me, you have been given opportunities, sometimes you have to work a little harder to ensure that each opportunity can also build character. To abstain from hardships altogether softens and weakens the mind and the body. Deprivation and desperation can be great motivators. So how better to replicate a form of hardship and abstinence than wilderness endurance?

It is the reason why ultra-marathons have become so popular. People crave a form of self-flagellation to help reaffirm what they have. It is so easy to become complacent. Unless you have perspective, you will become inured to what you have and instead start to lament what you don't have.

Absorbed in your own little world, it can sometimes be impossible to see beyond your four walls. The world is mighty. Full of people all fighting for something. Food. Shelter. Territory. Success. Fame. Human evolution is to aspire to something. Anything. The bigger house. The prettier or more handsome partner. The thinner waist. The bigger salary. The faster car. Society is built on a promise that if you work hard, you will work your way through the foundations of success. But is a big house, swimming pool and plastic surgery for enhanced boobs really success? Are these really the things we need? Or are they simply what we want, or are subliminally conditioned to aspire to?

On an expedition in the wilderness, none of the above will get you very far, not even the fast car. What we need to aspire to are possessing the tools for life and happiness. They are very simple qualities that strip us to the foundations of humanity. What's it all about? It's a biggie, but it can't be about wealth. What does wealth really get you? Now I'm not saying money doesn't help, but there is a tipping point at which too much becomes detrimental and unhelpful. It starts to have negative returns. We tend to be gluttonous and are never satisfied. We always take things too far.

We have become slaves to aesthetics. We care more about image and appearance than we do function. Take a look at all those 'perfect' human forms with ripped torsos, six-packs and bulging muscles on Instagram and ask them to take part in an endurance event, and I think you'd be surprised at how many fall at the first hurdle. Just as the body is an instrument not an ornament, life is about substance and depth, not superficiality.

The wilderness is the antidote. There is no façade or make-believe. It has a raw honesty that is almost simplistic in comparison to the complexities of modern life.

* * *

Back to the notion of proactive, voluntary hardship. Flagellation. Self-inflicted suffering, pain, misery and danger. Why do we do it? Why do I feel compelled to confront feelings and sentiments that most people spend their lives running away from? It's all down to learning.

If you have never known hunger or fatigue or thirst or fear or hypothermia or heatstroke, then you have never really tested the full potential of your own mind, body and soul. It's like having that fast car and never driving it fast.

I once took part in a BBC show called *Stars in Fast Cars* in which celebrities were given a selection of ridiculous vehicular challenges. One of them was to drive a Lamborghini around a race track as fast as you could, while towing a bathtub full of water without spilling any. The result was a race at 4 mph. Now I can say I have driven a Lamborghini around a race track – but have I really? Did I get to test it to the limits of what it was built for? No.

The same goes for humans. Very few of us ever really test ourselves. We take these incredible bodies and brains for granted. We might drag them to the gym or on a run every now and then, but who has worked their body like Mo Farah or Usain Bolt?

The reality is that modern life has surrounded us all in bubble-wrap and tissue paper. We treat ourselves like fragile porcelain that we are terrified of breaking or damaging.

That's not to say we are all lazy. There are some people who spend their lives pushing their limits. Some to great acclaim, others with no fanfare.

Have you ever dreamed of giving it all up and moving into the wilderness? How many of you have been on a holiday and wanted to remain in a hammock on the beach forever? There may be many of us who think it, but very few act on it because it's easier to follow the status quo.

I'll let you in on a secret. Climbing Everest and rowing the ocean were easier than my struggle to prove that I'm more than just a 'posh bloke'. The implication that I'm disconnected and out of touch. The reality is that I feel more connected because of the experiences and the people I have met. Experience has given me a unique understanding and a relatability that is borne through encounters, suffering and enduring.

I have experienced fear, loneliness, solitude, isolation, hunger and near-death.

I realise that a lot of my experiences were self-imposed, but it was never as simple as picking up a phone and calling for help. Helicopters aren't much good in the middle of the Atlantic. You are on your own. No amount of wealth or useful connections will help you. Nepotism is of no use. The ocean doesn't know what school you went to or what skin colour you have. *It doesn't care.* Because it doesn't matter. On your own, you have to dig far deeper than your wallet or contacts book. You go to a place that few of us really explore and you discover what gets you through it all. Resolve. Fortitude. Strength of character.

Let me be clear on one thing. There is great unfairness and inequality in the world. I have seen it everywhere and often, this growing chasm between the haves and have nots. I am a socialist at heart, except that I've seen how it destroys more than it unifies in practice, so I've never taken it further. We do need to fight inequality. But a great deal of that inequality is borne out of society. Consumerism and the seduction of marketeering creates tiers that you don't often encounter in the wilderness. You might collect a little more firewood or have more success with your tomatoes but ultimately you live a simple, relatively consistent life.

On an expedition resources are precious. You never waste anything or throw anything away. You reuse, recycle and ration everything you have. This is diametrically opposed to what consumerism encourages us to do.

Each time I return from the wilderness I do so with a promise to live a simpler life until I get seduced by a shiny new phone and abandon all hope.

We could in some ways use 'appearance' as a metaphor. When we think of people who live in the woods, the image that probably comes to mind is of a big beard, unkempt hair, a plaid shirt and ripped trousers. It's a cliché, but I've encountered it a lot. In Shoreditch, the hipsters call it 'Lumbersexual', but in the wilderness it's called 'I couldn't care less'.

The values are different in the wild too. A £200 pair of white Nike trainers aren't nearly as useful as a £5 pair of gumboots. A warm second-hand flannel shirt will be a hell of a lot more useful than a designer one. In the wilderness, we shift our values and things take on different levels of importance. What is of great value in New York, Edinburgh or Sydney will be of no value on Everest or in the Sahara, and vice versa.

It's this shift of values that is so important to experience and understand. It gives you perspective not just on yourself but also on other people. It makes you more understanding and considerate. It is too easy to live in a bubble of self-importance in which we are blinded to everything outside our own needs and our own little world. We become inward looking and fail to see beyond what 'we' think is important.

It's a tale of two worlds.

I straddle two worlds. It used to do my head in. I found it utterly confusing until I understood that it was a part of my apprenticeship. The learning I would need to grasp the reality of life.

The wilderness has shaped me and guided me. She has been both a guardian and also an antagonist. She has pushed me and moulded me. We have quarrelled and argued, loved and laughed.

There are so many lessons and parallels that we can take from our relationship with the wild. I prefer sunshine, but like everyone, I have my gloomy days when I am followed by a dark cloud that won't shift. We experience every human emotion in the wilderness. It may feel more intense and more immediate and our response more extreme because of the risks. Hope, love, joy, terror, frustration, happiness, and of course, loss. No story about emotions would ever be complete without the inclusion of loss.

I have loved and lost. Wept and mourned. Cried and missed. There is a saying in the sporting world that pain is weakness leaving the body, but there is no weakness in the pain of bereavement. When it comes to losing someone or something that you love dearly, it can be difficult to see through the fog of overwhelming, blinding bleakness that threatens to close over you forever. The notion that there is a way out can seem futile and hopeless, but there is a way. There is always a way.

As hard as it seems at the time, loss is the inevitable consequence of love. With the risk of love comes the risk of losing it. If you do not know loss then you probably haven't ever experienced love.

I want this book to be a powerful lesson in hope and happiness so I don't want to dwell on loss, but it is such a powerful instrument in directing us. It has the power to stop us in our tracks. A heady combination of emotions that remind us of our vulnerabilities and also of our mortality.

Loss, the heart-tearing emotion of losing something you love, can occur for many reasons. Dr Seuss once said, 'Don't cry because it's over, smile because it happened.' I love that quote. It is so full of wisdom, if only we could all live by it.

Loss has an ascending scale of despair from the heartache of unrequited love, to failed love, estranged family and death. Each one manifests itself in different ways and for different periods of time. I have experienced the whole range of loss, and each one has had a huge impact on me in its own unique, awful way. Friends who have left us too early, family who have died before I could kiss them farewell and dogs on whom I have depended.

It was the loss of our little boy Willem, stillborn at eight months, that left the largest hole. The shock, the disappointment and the confused emotion of missing a little boy who I never got a chance to meet.

I was in Canada for my late grandmother's 100th birthday when I got a call in the middle of the night. Marina had experienced a placental abruption. We had lost our son and Marina was also unlikely to survive.

The shock of the loss and the fear of losing Marina rocked me to my core. That journey halfway around the world to a tiny hospital in the Austrian Alps where Marina had been hospitalised remains one of the worst journeys of my life.

I have often considered death. I am often forced to prepare for my own and think about others'. When I head off on a faraway expedition, I must plan: the farewell letters hidden somewhere in the house and the will that is updated, but also the plan for what happens if a loved one falls sick or worse. What if I am halfway across the Atlantic with no way of speeding up my journey? Am I told straightaway, or do they wait until I arrive? It is a morbid truth of expeditions and faraway journeys. Do you want to be told of someone's death while you can do nothing to hasten your return to civilisation, or would it be more sensible to be told at the end? I have faced this conundrum many times and there is rarely a consistency in my response. The point is that loss is an inevitable part of humanity.

Losing little Willem while I was on the other side of the world was wretched and life-changing. I felt I had lost control. No capsize or crevasse field had prepared me for that awful journey home and the unknown of whether Marina's life would be saved.

We were so lucky that Marina eventually made a full recovery. But the relief that she lived was counterbalanced by the loss of our little Willem.

As I walked into the hospital room, a kindly nurse handed me my little stillborn son. He was perfect. He looked like he was sleeping. We named him. We had to organise his cremation, the repatriation of his ashes, a funeral, a memorial stone and then the social maze of how to tell people. How do you do that without ruining someone else's day? 'Ooh, what did you have?' people would say upon seeing Marina's flat stomach after eight months with a bump.

I don't need to dwell on what was a very painful experience, but it left me with not only a profound sense of loss but also a bout of anxiety, something I had never experienced before. It was debilitating. I found it hard to concentrate and became socially introverted. Loss, the wretchedness of which touches us all, is a human leveller.

But how could I force myself to regain control of the situation? My therapy was, of course, the wilderness.

The afternoon after holding my little stillborn son, I picked up Ludo and Iona who were being looked after by my parents-in-law and we headed to the mountains. We went all the way to the top of the Gaisberg mountain that overlooks Salzburg. The clouds were below us and the summer sun was still high in the sky. It cast an ethereal light on my children as they ran in the meadow. That moment is indelibly marked in my mind; that despite the loss, it was the celebration of what I still had. I walked those mountains for a week, to build the strength for the shared journey back from the darkness.

Loss and recovery. I have always believed that life needs texture and contours. Ups and downs. Highs and lows. If we live it on a perpetual line of idyllic happiness then we lose sight of what we have.

* * *

When I think back to my childhood and my sense of failure and inferiority, I recall the battles I had to prove myself. There were the sporting jocks and those who were brilliant academics, and then there were the rest of us ... the geeks who were yet to find their roles in life. But here's the thing: those friends who were brilliant at rugby or those who got A*s without even thinking, or those who were the most beautiful or handsome, rested on the laurels of what life had given them. They didn't need to try hard in life. On that towering line of consistent perfection, they didn't need to fight.

My life has been a personal inner battle to rise above the flatline of mediocrity and prove to myself and to others that there was more to me. I wanted to show that I was worth more than a place in a sports team or a high exam grade. The early failure fostered the spirit to fight. And it's that spirit that has led me on a journey to where I am today. Without the early failure and imperfections I would never have 'tried'. There would have been no oceans to cross or mountains to climb. How different would my life have been? Our experiences are what make us the people we are. The losses, the gains, the loves, the failures, the seemingly insurmountable barriers.

I am 46 at the time of writing this book. I don't feel old and I don't fear death. I am losing years from my life but I would prefer to focus on what I still have ahead of me, rather than dwell on the years that are escaping me. You are only as old as you allow yourself to feel, and for me I still feel like an 18-year-old.

When I turned 40, something happened and I became comfortable with myself. At ease with my body. My appearance. Me. As David Bowie once said, 'Ageing is an extraordinary process whereby you become the person you always should have been.' Suddenly, I no longer felt a need to conform to how others wanted or expected me to look or be. For 40 years I had worried about my appearance. As a teenager, I had hated the spots that pock-marked my face. In my twenties, I worried about my big nose. In my thirties, I worried that everyone had bigger biceps and pecs than me. But my forties were different. I finally accepted me for who I am, not trying to look like or behave like society expects. My body and personality had been shaped. I bear the scars of my experiences and like that shattered vase or teapot, each repair is a reminder of the opportunities I have had to rebuild myself.

The wilderness is itself experiencing its own heightened loss of habitat, animals, wildlife, glaciers, rivers, forests, coastlines ... and more.

We are all responsible. Humanity and the parasite of consumerism is destroying our planet at record pace. We have left our mark like scars on a body.

We haven't evolved. We have devolved. We may have made our lives 'easier' but at what cost to our physical and mental well-being? Is it any wonder we are facing a mental health epidemic? We may have invented fast food, fast cars, fast fashion and fast escalators, but we have also had to fill our lives with diets, vitamins, alternative medicines and regular visits to our GP. We have become slaves to speed and efficiency to the detriment of our health.

Society sells us the 'dream'. Go to school. Work hard. Pass exams. Get into university. Get a first. Get a job. Take a mortgage.

Buy a house. Spend your life paying off debt. Die. Of course this is a brief overview without the nuances, but you get my point.

That is not to say it isn't pierced with moments of happiness, but they become the treat rather than the main event. They are the pudding to the dinner of drudgery. The hamster wheel of life. Every now and then you step off it, when you take a holiday or get blindingly drunk on a Friday night, but you better be quick to get back on, or you'll miss your turn and end up with nothing. We have become institutionalised to the 'system'. Now I'm not extolling some hippy shit. I'm no guru. To many I am a privileged posh bloke who has exploited opportunity, but I have seen another way and it doesn't have to involve following a plan or a prescription.

I believe happiness is the ultimate goal in life, so how do we attain a relatively good level of happiness in our lives? It can't be the rollercoaster line of euphoria followed by despair followed by euphoria again, as that isn't sustainable. We need to find a constant. This constant is the key to life. How do we achieve it? Well, the answer is that it is unique and bespoke to each and every one of us. No two are alike.

If you live with nothing, no money, no food, no job and no prospects, then the leap to happiness might be a slice of bread or a safe place to sleep for the night. Happiness is relative. If you are a billionaire with multiple mansions, private jets and expensive cars, then it might be something as simple as an honest friendship that brings happiness. It's all relative to what you have or don't have. Now I realise that for most of us, money is the key to security, which in turn brings an element of happiness. When life becomes more than a struggle for food, shelter and safety, you can have a sense of reassured complacency.

As a child, happiness for me was always when I stepped out of the system. When I stepped off the hamster wheel. Holidays in Canada. Walking the dogs in the park. Watching Dad operate in his vet clinic. Weekends on the farm in Sussex. Unhappiness came from education. Exams and the pressure to conform to a system that wasn't suited to me. I failed my exams and exams failed me. I have been very vocal about my disdain for both exams and homework. They both brought unhappiness that was hard to shake from my childhood. They brought an anxiety and nurtured a lack of confidence. My parents sent me for extra tuition. I did Easter crammer courses at great expense during the spring holidays to help me with my A levels and GCSEs, to no avail. The result was that I felt a guilt that my parents had worked so hard to pay for all that tuition and I had failed them both.

The wilderness is the antidote to the burden we place on ourselves. The wilderness is like an organic classroom, which nurtures skills for life rather than skills for the city.

I have always hated the judgement of society. We make decisions about people based on appearance, hair colour, height, skin tone, sex, complexion, beauty, intelligence, provenance, humour ... the list goes on. I understand that we need to differentiate people. The world would be incredibly boring if we were all cookie cutter shapes. We need to be different and celebrate uniqueness, but we shouldn't be so judgemental.

Intolerance and illiberalism have created a complex society that can be difficult to navigate as a 46-year-old, let alone a youngster setting out in this world. For someone in the public eye, I am incredibly thin-skinned. I hate criticism and I take it to heart. People are so quick to point out shortcomings rather than celebrate achievement. People will always find an excuse to bring you down. You can spend a lifetime building your reputation, only to watch it crashing down with one misjudged tweet or ill-advised statement.

The wilderness is not like that. She is far more forgiving. That's not to say she doesn't have the power to take a life with a single avalanche, or flood or drought, but she is far less judgemental than society. I realise I am anthropomorphising a whole environment, but the wilderness really can be the cure to so many of our ills, if only we were a little more open-minded.

Modern life is about overindulgence. Everything needs to be bigger, better and faster. The result is a lifestyle in which we don't have time to repair or rebuild. We simply discard and replace. In some ways, people have become like that too. We have become so intolerant and rash that we simply discard people after one digression rather than give them the time to redeem themselves.

Can you imagine the pressures this brings to youngsters heading into the world? We like to think of ourselves as progressives but in fact we have been handcuffed to a system that relies on aggressive competition. When did life become a race? A competition? I have always embraced it as a marathon, not a sprint. That's not to say there aren't winners and losers but you have the best part of a hundred years to make your mark, not 15 minutes.

We live now in an age where your every waking moment can be broadcasted and published for all the world to see. Instagram, TikTok, Twitter and Facebook have made many of us global stars, featuring in our very own version of Hollywood's *The Truman*

Show. The result is that we have never before come under such scrutiny. If I sit back now and think about each of the words I place on this page, I am liable to suffer a panic-induced anxiety attack based on my catastrophising and how it might be interpreted. That is where we are. A now confident, relatively successful, 46-year-old father of two who has done some dangerous shit, still feels worried about reaction, interpretation and image.

Perhaps it is no surprise I feel such ease, comfort and happiness when I am in the wilderness. It is the place where I can switch off from the system. It brings a sense of belonging. I breathe easier. I am calmer. I walk more slowly. The pace of life changes and I feel my anxiety subsiding.

Try it. Lie on the floor of a forest or in the woods and stare up at the canopy above. Winter or summer, lie there and embrace nature. There are plenty of cultures who have prescribed this form of forest bathing as a remedy to the ills of modern society. My head is clearer and my mind becomes laser like. Now this is just when I am in the wilderness, but imagine now what happens when you add 'endurance'. By endurance, I refer to suffering and hardship born through adversity.

You can be in the wilderness but you can also become a part of that wilderness by relinquishing all comforts and immersing physically and mentally.

Physically we can go for a run, eat chocolate or have sex. We produce endorphins that bring short-term happiness and pleasure. But how do we prolong this? How do we clear our minds that are full of the clutter that we absorb from everyday life?

Fill your mind with task and goals and suddenly what the Kardashians ate for breakfast or what Trump just tweeted become comparatively inconsequential. I'm not saying that what Trump has to tweet isn't important, but we don't need to clog up so much of our already bursting brains with the minutiae of other people's lives and what they had for breakfast. We need to focus on the here and the now.

There is no better way to do this than to endure. To suffer. To sweat. To go hungry and thirsty. To feel fear. Real fear. To stare death in its face.

The thing about voluntary risk is that you only have yourself to blame. If something goes wrong, you can't go knocking on the door of the nearest no-win, no-fee 'tripped on a pavement' lawyer. You stand or fall (or trip) based on your own actions.

Society loves to blame. The result has been the loss of 'risk' in most of our lives as no one wants the culpability. Blame culture has hampered ambition and opportunity. I sometimes think people are faster to blame than they are to compliment or congratulate.

What kind of society have we created when people are making decisions on what people will think rather than what is the best decision? It's messed up. Once again in the wilderness, decision making does not rest on image or interpretation. It is done on life and death. Sustainable or unsustainable. There is a pure binary nature to it. It is that simple, organic, slow pace that makes the wilderness such a therapeutic place to be.

I have experienced the wilderness in her many manifestations and wearing many uniforms and each one has given me a new understanding about myself. An understanding that I never had time to digest when my mind was polluted with the weight of expectation from society.

Expectation has an underrated toxicity. There are so many things that we are 'expected' to do in life. Anyone who chooses otherwise is 'eccentric', 'alternative' or 'mad'. We are back to the shepherd or the sheep metaphor. Isn't it crazy that we are told to be 'sheep', to conform and fit in? As children, we crave 'sameness'. Those with insecurities don't like to stand out as it shines a spotlight and draws attention. But over the years, I have learned to understand the importance of individuality. Take a look at the flora and fauna. It is different in each environment and landscape, uniquely adapted.

Homogeneity. We are living in an increasingly homogeneous era. Cars and houses and even people are beginning to look the same. Bizarrely, in an age when we should be celebrating uniqueness, we seem to be heading to vanilla-like sameness. Perfect teeth. Big boobs, bulging biceps and a six-pack have become the accepted norm. It is with this backdrop that I have been able to contextualise my own experiences in the wild.

This brings me back to the Japanese art form kintsugi, to that broken vase repaired with golden threading, whereby the repairs become a beautiful art form, and the vase, with its spider's web of repairs, is enhanced. I was once a broken vase. It's taken me many years to piece myself back together. Today I am no masterpiece, and I'm certainly not precious or valuable like the kintsugi vases, but I wear my scars with pride and I am stronger.

About a decade ago, in my mid-thirties, I got my first tattoo. It was a drunken one, on my back while training for the South Pole by pulling tyres in Devon. Marina went crazy. She called me selfish and she still makes a retching noise every time she sees it. She hates tattoos and calls mine my 'midlife crisis'. I have since had a further three. I used to hate tattoos, so what happened? Why did I get them?

I suppose the answer is confidence.

Each one of my tattoos is a considered reminder of who I am, what I have done, where I have been. If I am ever down or feeling vulnerable I look at them and am reminded of tougher times. They are a sign of the confidence in my inner self. A reflection of me. I didn't do them for others to admire, but for me.

I suppose Marina is right. That does make me selfish, but it's not the midlife crisis. It's easy to write off such behaviour as midlife doom, but I beg to differ. I think they are marks of my new-found self-confidence. This is me. Take me or leave me. Midlife is just that. A moment to reflect. A little like the glass half full, I think life has only just got going.

I've worked my way through the bumpy bits of insecurity and I've found my inner confidence. Midlife doesn't have to be a crisis. I think it's a celebration. You've successfully navigated the turbulent waters of the first half and can finally embark on the second part of your life, armed with knowledge, confidence and experience.

For me, I've reached a point in my life when I can do things, and try things, with conviction. When I climbed Everest, there were some who interpreted it as a midlife crisis, but I disagree. I don't think I could have done it in my earlier years. I simply wasn't equipped to deal with the mental and physical toil. With maturity comes an ability to understand your own physical and mental well-being. Armed with the experience of success and failure, we are better prepared than we were in our youth when puppy-like naive enthusiasm gets you into all sorts of trouble.

I am now 46. I don't feel middle-aged but everyone keeps reminding me that I am. I feel no crisis. I see the gathering grey whiskers in my beard and notice the lines becoming more marked on my face, but I don't feel the need to hide the marks of time. The lines, the scars and the tattoos are proof that I have lived. That I have embraced life.

I don't feel old and I don't fear getting older. We all get older. Fact. The question is, do we need to change our behaviour as we grow older? Are we bound to society's expectations? Why should we be? I hope that I am still climbing mountains when I'm 80, not to look like I'm still a youth but because it makes me happy.

The wilderness fosters a relationship that is rarely selfish or wasteful. The competition between birds is not about who can build the bigger nest and waterfalls don't compete for the longest drop. That's not to say there isn't a competition for resources but it is largely based on need rather than want. The crux of this is happiness gained through the provision of life's necessities.

Happiness. That is what the wilderness means to me. Happiness is a place between too much and too little. Underrated and often overlooked, happiness is the key to almost everything. Surely it is the meaning of life. To experience happiness is the ultimate emotion. Of course it needs contours and fluctuations for maximum effect, but there is no reason why we can't all attain a baseline of happiness and fulfilment.

I am not so naive as to conclude that we are all armed with the necessary ingredients for that happiness. Conflict, division, racism, sexism, war, hate, hunger, poverty have made the world an unequal, unfair place. Society and humanity foster the ills that make us unhappy. The wilderness can be the medicine and the vaccine. It strips us of the things that make us unhappy.

We have lost our way and our moral compass. In this divided world, it can be hard to know what is true and what is false, what is right and what is wrong. In the polarisation of the right and the left, we are forced to walk a tightrope, one digression from which will send us tumbling into the abyss of outrage. We have become conformists. But it doesn't have to be like this. Be the shepherd not the sheep.

While urban 'society' remains obsessed with teaching children maths, spelling and reading, I would prefer my kids to know how to use a knife, start a fire and skin a rabbit. Bushcraft and wilderness skills have been lost, to the detriment of society. While plenty of schools thankfully offer bushcraft to their children, it still isn't part of the curriculum. Schools are still 'boxes' built to tick boxes.

My time in the wilderness, living with people in the wild and on expeditions, has taught me to respect nature. The enforced rationing of expedition life teaches you to value commodities and resources. Every drop of water, morsel of food and last piece of firewood can be the difference between life and death. There is no place for gluttony in the wilderness. Excess is death.

Governments rely on societal conformity. Without our taxes, they are screwed. Many of those I have visited have struggled to break free from the long arm of the government, which doesn't want them to 'leave' the monetary grid. But beyond the collapsing world order, the chaos of Brexit and the divisiveness of Donald Trump is a world of organic simplicity. While we are obsessed with clean eating, organic food, veganism, exercise, mental well-being and yoga, we have failed to address the most harmful, stressful aspects of life. Where we live. The urban city. More of us could adopt a greener, cleaner life that is gentler on our environment. I'm not suggesting that we all move to the wilderness; that would be disastrous, but we can all become more self-sufficient. More resourceful and less wasteful. We need to wean ourselves off materialism. We have become slaves to the seduction of big companies wanting to sell us the newest, most advanced, most on-trend products, when what we already have will do. We need to look back to look forward. Make do and mend.

The wilderness in all her forms teaches us to respect and value nature. In an increasingly urbanised society, so many of us have become disconnected from our flora and fauna. The wilderness is a place in exotic documentaries. A theme park to visit fleetingly. While the world's off-gridders might thankfully still be a minority worthy of making TV shows about, I can't help but think they are ahead of us. We may be technologically advanced but have we socially regressed? In reverse Darwinism, perhaps the wild folk will be the ultimate survivors?

The health benefits of the wilderness are only now being celebrated. The wilderness strips us back to our basic instincts. It helps to recalibrate. Each time I return from periods in the wild, I feel calmer, freer, wiser, and then I become seduced by that great seductress that is society and all her promises.

Every one of those wild folk I have visited has forfeited the assurances and comforts of society for a grittier, hand-to-mouth existence. Their lives are rarely easier, but boy are they happier. They are unified in their collective happiness. How many of us dream of embracing a simpler life? How many of us spend a lifetime asking what if? Well, those brave souls have done it. They have lived their dream. Disconnected from the material world, they have become free.

While many in the developing world still head to the cities for opportunities and hope in the pursuit of their dreams, many of us from the developed world are going the other way. Moving back into the wilderness. As my own children grow up, I will be showing them that there is another way. Break free from normality and they can be truly free.

As the world continues to grow, we will not only place untold pressures on our natural world, but also on ourselves. With population growth, as the competition for space, and jobs, and companionship becomes more ferocious, so will our selfishness and stubbornness. We have a blind spot to our own health and well-being. We are so busy trying to accumulate and consume that we have lost the ability to be rational and realistic about what we really need. So we are consuming our very health and well-being in the pursuit of a false promise.

We have become blinded by the pound, the euro and the dollar, when really all we need is water, air, plants, shelter and nature. Simplicity is key to happiness.

* * *

Like a broken vase or teapot, I have slowly pieced myself back together, and while the lines of repair haven't always been perfect, some of the repairs are more solid than others; although the vase remains fragile, it is also an awful lot stronger.

We are all fragile. We are all vulnerable. None of us is perfect: perfection is the enemy of progress. The search for perfection will only impede improvement. The wilderness has toughened me and strengthened me, both in body and soul. My spirit has been lifted and my confidence slowly rebuilt. I am a humbler person, more thoughtful and considerate, and I'd like to think I'm also kinder. The wilderness has been my healer, my companion and my guide.

I may have failed school exams, but I have finally graduated the school of life. I have not just survived but thrived. This wasn't a battle I won, but a symbiotic relationship I formed. One that fosters generosity rather than greed.

The importance of human spirit should never be underestimated. With spirit comes confidence and hope and happiness. Just imagine if we put as much emphasis on spirit as we do on exams. Imagine if we focused on building the inner confidence to be who we are, not what others want or expect us to be. Imagine if we lived in a world where there were more shepherds than sheep. Happiness is there; all we need to do is take away the things that make us sad and it will be ours.

I'm aware of the irony in the old saying, 'too many cooks spoil the broth', but have you ever asked yourself to consider how delicious the broth might be if those cooks worked together with their collaborative individuality?

The wilderness has nurtured individuality that has fostered the confidence of spirit. It was the much-respected nineteenth-century mountaineer and environmental campaigner John Muir, who once said that 'the clearest way into the universe is through a forest wilderness'.

The wilderness encourages the freedom of individuality and expression. It equips you with the skills and the resourcefulness and the respect to be kinder and humbler and happier.

In my case, the broken vase has finally been repaired; the repairs, like the scars on my body, are a proud testimony to the sweat, the tears, the triumph and the tragedy that got me to this point. The wilderness has given me the confidence to be who I am, not what others want or expect me to be.

It is so easy to lose perspective when you are in danger of arrogant complacency or fearful anguish. Don't lose yourself to the enormity of the situation. I have often felt overwhelmed and dwarfed by the scale of the challenge ahead. Imagination is a fearful emotion and once you lose control of your mind you are in danger of conceding to the panic within.

Be adaptable, resourceful and creative with what you have, rather than always worrying about what you don't have.

We have become so used to a fast daily routine that it can feel strange and alien when we are forced to slow down. The period of the Covid-19 pandemic has given us the opportunity to step off the perpetually turning wheel and have a moment to reflect and consider.

When you are in the middle of the 'storm' it can often feel like it will never end, but it always does. The blue skies always return eventually. It is one of the few certainties in an uncertain world. The key is to batten down the hatches and see out the bad times. Don't let negativity take over. There are some things we can only learn in bad weather. Storms encourage trees to take deeper roots. They can strengthen us as long as we don't give in to them entirely.

Be who you are, not what others want you to be. People will judge you but don't let that judgement define you. Don't let failure defeat you. Happiness is about letting go of what you think your life is meant to look like and embracing what you have rather than what you want. If you want nothing then you have everything. Be inspired by that thought.

To be happy, you just need to be a little more wild.

EPILOGUE

Autumn 2020

A treehouse in Oxfordshire

Dear Wilderness,
I have missed you.

One hundred days isolated from your isolation has given me time to reflect. Sometimes you don't appreciate what you have until it is gone. This period of lockdown has reminded me that we must be a part of you, not apart from you.

You have been my guide and mentor and you have helped me navigate the complexities of modern life.

You have taught me to be resourceful and considerate. Your oceans have taught me resilience and fortitude. Your mountains have helped me overcome my fears. Your rainforests have taught me to be less controlling and to concede to the uncontrollable rather than exhaust myself fighting it. Your deserts have healed the spirit and soul. Your ice caps have taught me how to cope with the solitude of isolation and remoteness. But most importantly you have taught me the most valuable lessons for life.

This is what I have learned:

Be kind.

Think of others before yourself.

Humanity relies on the collective spirit.

It is easier to love than to hate.

Anger doesn't solve anything but it can destroy everything.

The most important thing in life is happiness.

Fast cars, fast food and fast fashion may bring you brief pleasure, but slow things will bring you a lifetime of happiness.

Do less and be more.

If you are always racing to the next thing, then what happens to the present?

We are more likely to regret the opportunities we didn't take.

Rain may be miserable but we need it for food and water, and after the rain comes the sunshine.

Try to forget the mistakes and remember the lessons we learned from them.

If you want nothing, you have everything.

Don't get confused by consumerism. Do you need it or want it?

Positivity is partly a state of the mind. It's always there, it's just that some days you'll need to look a little harder.

Believe you can and you are halfway there.

Impossible is not a fact. It's just somebody's opinion.

You can't control the wind so adjust your sails and work with it, not against it.

The key to adapting is to accept your circumstances. Don't battle them.

Strength and fortitude will get you far in life. Never doubt yourself. You never know how strong you are until strength is the only choice you have.

The key to success is to focus on goals, not fixate on obstacles.

Resilience is the art of overcoming the unexpected.

Tough times don't last forever, but tough people do.

Try and be selfless, never selfish.

Thank you for healing me and showing me the strength and spirit to be me – the real me, not the person society wants me to be.

With love and thanks,
Ben

ACKNOWLEDGEMENTS

Thanks to everyone who has helped me find the wilderness path.

Thanks to Myles Archibald, Hazel Eriksson and Helen Upton at William Collins for believing in this book. And thanks to Martin Toseland and Tom Whiting for their keen editorial eyes.

INDEX

BF indicates Ben Fogle.